W9-CHP-581

ACKNOWLEDGEMENTS

We would like to acknowledge and thank the following people. Without their efforts, our project would remain an idea in our heads instead of a book in your hands.

Theresa DiGeronimo for her tireless and selfless editorial work and professional guidance throughout the process.

Jason Chen for his tremendous artwork that graces the book cover.

Colleen DiGeronimo for sharing her peerless talent on the book cover's graphic design.

EXTREME OPERATIONAL EXCELLENCE

Applying the US Nuclear Submarine
Culture to Your Organization

MATTHEW DIGERONIMO
BOB KOONCE

outskirts
press

MENTORS

Throughout our careers, we were blessed in many ways. Perhaps most significantly, were the mentors that selflessly devoted themselves to our professional and personal development. This book along with our professional achievements would not have been possible without their mentorship and friendship.

Matt DiGeronimo's Mentors:
Admiral Kurt W. Tidd, USN
Vice Admiral Robert P. Burke, USN
Captain Shan Byrne, USN
Captain Paul Dinius, USN
Captain James Doody, USN, Retired
Captain Charles Melcher, USN, Retired
Captain John Peterson, USN, Retired
Captain Kenneth Swan, USN, Retired
Commander David Cox, USN
Commander Dennis Klein, USN, Retired
Commander Robert Koonce, USN, Retired
Dr. Thomas Stickle

Bob Koonce's Mentors:
Rear Admiral Jeff Jablon, USN
Rear Admiral Doug McAneny, USN, Retired
Captain Bob Brennan, USN, Retired
Captain Mark Breor, USN, Retired
Captain Ray Lincoln, USN, Retired
Lieutenant Commander Matt DiGeronimo, USN, Retired
Lieutenant Commander Steve Smith, USN, Retired
Master Chief Tim Hella, USN, Retired
Master Chief Tim Preabt, USN
Mr. Bob Carritte
Mr. Jim Connell
Mr. Andrei Perumal

ABOUT THE AUTHORS

Matthew DiGeronimo was born and raised in Hawthorne, NJ. He is a summa cum laude graduate of Rensselaer Polytechnic Institute (BA Nuclear Engineering), University of Connecticut (MBA), and a PhD candidate (Business Psychology) at the Chicago School of Professional Psychology.

In the Navy, Matthew served aboard the USS HAMPTON (SSN-767) and the USS KEY WEST (SSN-722) as the Engineer Officer. Additionally, Matthew was the Lead Submarine Tactics Instructor at Navy Submarine School and a Junior Member of the Nuclear Propulsion Examination Board. In 2003, he volunteered to serve on the staff of COMMANDER MIDDLE EAST FORCE in Bahrain during which time he contributed to Operation Iraqi Freedom and Operation Enduring Freedom in a variety of capacities.

Matthew's post-Navy experience is highlighted by his successful turnaround of three businesses in diverse industries: a mergers & acquisitions company, a commercial real estate and property management company, and a district energy power plant.

ABOUT THE AUTHORS

Bob Koonce was born and raised in Blue Mound, Illinois. He is a graduate of the University of Illinois in Electrical Engineering and earned an MBA from Northwestern University as well as a Masters in National Security and Strategic Studies from the Naval War College.

Bob served on five nuclear submarines during his twenty years in the Navy including Command of USS KEY WEST (SSN 722) where the ship earned the 2009 Battle Efficiency "E" for excellence in Submarine Squadron Three. Bob also served as both a junior and senior member of the Pacific Fleet Nuclear Propulsion Examining Board and mentored Midshipmen at Northwestern University.

Following retirement from the Navy in 2011, Bob has worked as a consultant and manager in the power industry. Bob frequently speaks and consults on Operational Excellence. He is passionate about helping leaders develop their teams into high reliability organizations and achieve excellence.

DEDICATIONS

To Mom, your life's work inspires me
To Aimee, your love motivates me
To God, your grace and mercy saves me.
Philippians 4:13

- Matt DiGeronimo

To my parents who taught me integrity and hard work.
To Rita and our wonderful kids for supporting me in everything.
To my Lord and Savior, Jesus Christ, for forgiving me.
Proverbs 3:5

- Bob Koonce

To all of our shipmates, we are extraordinarily honored
and grateful to have served with you. Our pride in our
shared service and common bonds is the heart and soul
of this book. To each of you - thank you.

Dolphin Code 34

TABLE OF CONTENTS

INTRODUCTION

We are extraordinarily proud to call ourselves United States Submariners. We are proud of the community's heritage and tradition of operational excellence. We are proud to have served with some of the best, brightest, and bravest sailors in the fleet. We are proud of our individual and unit accomplishments that we enjoyed during our careers. However, our pride is not why we wrote this book.

This book is not a memoir. This book is not about "us." This book is not a "look what we did" book. In fact, we have an imperative message to all of our shipmates and to all of the readers of this book. The content of this book was derived from our shortcomings as leaders. There were cringe-worthy moments while writing the book when we realized that a particular piece of guidance was one that we had wished we had taken to heart while we were serving. Using our shortcomings as fuel for this book is consistent with the message of the book. We all look and feel good when we perform well, but the difference between *average* and *excellent* is how we handle our less-than-perfect performances. Many great organizations, knowingly or not, sweep many of their imperfections underneath the rug. One of the most amazing elements of the Nuclear Submarine culture is its self-enforced refusal to sweep problems under the rug. For decades the submarine culture has recognized the criticality of squeezing out every ounce of lessons learned from imperfect performances.

Between the two of us we have experienced many cultures outside of the Navy since hanging up our uniforms—small businesses, privately held companies, and large publicly traded corporations include international companies in a wide range of industries including retail, mergers and acquisitions, communications, entertainment, utilities, consulting, engineering design, and construction. During this time, we frequently exchanged notes, and we noticed that there was a common theme throughout our experiences. That theme centered on the belief that each organization could significantly improve its operations through the use of the fundamental components of the Nuclear Submarine culture.

We wrote this book in an effort to share the culture of operational excellence in the United States Navy's Nuclear Submarine community because we know that the elements that underpin the success of the Nuclear Submarine community are all transferable, in various forms, to the organizations that we have observed.

We are patriots, men of faith, leaders, and students of leadership, but we also consider ourselves quite ordinary in comparison to those that we served with in the Nuclear Submarine community. We each know many men whose ability to lead, demonstrate and articulate the lessons discussed in this book far surpass our own. It is because of our admiration of and dedication to these men, and all of our shipmates, that we were able to write this book.

From the authorization of the USS Nautilus (SSN-571) in 1951 to the emergence of today's state of the art Virginia class nuclear submarines, the Nuclear Submarine community has accomplished more than any book could capture. This book does not attempt to capture those accomplishments. This book is bereft of ego and bravado—our own and of the Nuclear Submarine community. In fact, it is important for the reader to understand that the submarine anecdotes shared in this book tend to provide examples of events that were not performed perfectly. These anecdotes are anomalies that we are using to highlight how best to leverage the imperfections of an organization into operational excellence. We also share them because operational excellence does not imply that an organization is devoid of mistakes. Speaking most literally, organizations pursue operational excellence—they do not achieve it. An organization may approach operational excellence asymptotically, but the closer it gets, the more aware it becomes of its own imperfections. It is this awareness, coupled with the action taken in response, that represents a large portion of what we call *operational excellence.*

The idea of being one's own worst critic through brutally honest self-evaluations originated from the philosophies of the Father of the Nuclear Navy, Admiral Hyman G. Rickover. We are going to discuss more about Admiral Rickover, but first allow us to set the stage for his vision.

Prior to the use of nuclear power for propulsion and electric generation, submarines were diesel powered. To describe these submarines as "diesel powered" is wholly consistent with submarine vernacular—these boats were, and still are, referred to as "diesel boats"—but the expression is slightly misleading. "Diesel boats" were more literally "battery boats." It was necessary to partially surface the submarine to run the diesel engine (which cannot be operated while a submarine is fully submerged) in order to charge the the submarine's batteries. The noise of a running diesel engine can be detected by even rudimentary sonar systems from miles away, not to mention the visual overtness of a partially submerged submarine charging its battery. Because of these two elements, remaining covert was challenging, to say the least, for diesel boats. Once the submarine had a charged battery, it could then fully submerge and conduct the covert operations it was assigned, powered by its battery. However, the battery's charge was limited, especially if the submarine traveled at speeds beyond a snail's crawl. Therefore, submerged operations were extraordinarily limited in duration.

Admiral Rickover's Naval career was like none others. He was commissioned after graduating from the Naval Academy in 1922. He retired as a four-star admiral, 60 years later, in 1982. Early in his career, he served on several diesel boat submarines, battleships, and minesweepers - including a three-month stint as the Commanding Officer of the USS Finch, a minesweeper before transferring to the Engineering Duty Officer community. During World War II, he led Navy Department's Bureau of Ships, after his instrumental role in repairing the Battleship, USS California, from the damage it sustained during the attack on Pearl Harbor. The seed for the future of submarine warfare was planted when he was assigned to the Manhattan Project in 1946 where the emphasis had shifted from harnessing atomic power for weaponry to harnessing atomic power for electric generation. During his assignment to Oak Ridge, Tennessee as a Navy representative to the Manhattan Project, he developed a vision.

He envisioned using nuclear power to provide propulsion and electrical power for submarines. This would allow submarines to remain submerged indefinitely for all practical purposes. (To this day, the length of a Nuclear Submarine's submerged operations is limited only by the amount of food the submarine can carry.) Of course, Admiral Rickover's vision was not easily converted to a reality. There were a few minor details—designing a machine that rivals the space shuttle in its complexity, building such a machine, training the crew and supervisors to operate this machine with the highest standards of operational excellence, and obtaining the Navy's and Congress' approval to do so. Remarkably, through Admiral Rickover's relentless determination and commitment to operational excellence, his vision became a reality.

Admiral Rickover was truly a man before his time. He was obsessed with quality assurance and operational excellence when most of modern industry was several steps behind him. His leadership style and philosophy was so unique that it remains the thread woven into the fabric of the Nuclear Submarine's culture to this day. The unique nature of this philosophy is best expressed directly from the source. As such, here are excerpts from a speech[1] that Admiral Rickover gave at Columbia University in 1981. It is uncommon to provide such a large portion of a speech transcript as this; however, the density of information and high-level thought in this speech is too remarkable not to share. Further, you are reading the seeds of the U.S. Nuclear Navy's journey towards Operational Excellence which sets the stage well for the remainder of the book.

> *Human experience shows that people, not organizations or management systems, get things done. For this reason, subordinates must be given authority and responsibility early in their careers. In this way they develop quickly and can help the manager do his work. The manager, of course,*

1   Rickover, H. G. (1981, November 5). Doing a job. [Speech made at the 1981 Egleston Medal Award Dinner, Columbia University School of Engineering and Applied Science, New York.] In Hearing Before the Joint Economic Committee Congress of the United States, 97th Congress, Second Session. *Selected Congressional Testimony and Speeches by Adm. H. G. Rickover, 1953–81* (pp. 756–773). Retrieved from U.S. Congress Joint Economic Committee website: http://www.jec.senate.gov/public/index.cfm/1982/12/report-ada7008e-9768-4608-b47c-b1dd2af1e8d4.

remains ultimately responsible and must accept the blame if subordinates make mistakes.

As subordinates develop, work should be constantly added so that no one can finish his job. This serves as a prod and a challenge. It brings out their capabilities and frees the manager to assume added responsibilities. As members of the organization become capable of assuming new and more difficult duties, they develop pride in doing the job well. This attitude soon permeates the entire organization.

When doing a job—any job—one must feel that he owns it, and act as though he will remain in the job forever. He must look after his work just as conscientiously, as though it were his own business and his own money. If he feels he is only a temporary custodian, or that the job is just a stepping stone to a higher position, his actions will not take into account the long-term interests of the organization. His lack of commitment to the present job will be perceived by those who work for him, and they, likewise, will tend not to care. Too many spend their entire working lives looking for their next job. When one feels he owns his present job and acts that way, he need have no concern about his next job.

In accepting responsibility for a job, a person must get directly involved. Every manager has a personal responsibility not only to find problems but to correct them. This responsibility comes before all other obligations, before personal ambition or comfort.

A major flaw in our system of government, and even in industry, is the latitude allowed to do less than is necessary.

Too often officials are willing to accept and adapt to situations they know to be wrong. The tendency is to downplay problems instead of actively trying to correct them. Recognizing this, many subordinates give up, contain their views within themselves, and wait for others to take action. When this happens, the manager is deprived of the experience and ideas of subordinates who generally are more knowledgeable than he in their particular areas.

Unless the individual truly responsible can be identified when something goes wrong, no one has really been responsible. With the advent of modern management theories, it is becoming common for organizations to deal with problems in a collective manner, by dividing programs into subprograms, with no one left responsible for the entire effort. There is also the tendency to establish more and more levels of management, on the theory that this gives better control. These are but different forms of shared responsibility, which easily lead to no one being responsible—a problem that often inheres in large corporations as well as in the Defense Department.

A good manager must have unshakeable determination and tenacity. Deciding what needs to be done is easy, getting it done is more difficult. Good ideas are not adopted automatically. They must be driven into practice with courageous impatience. Once implemented they can be easily overturned or subverted through apathy or lack of follow-up, so a continuous effort is required. Too often, important problems are recognized but no one is willing to sustain the effort needed to solve them.

The man in charge must concern himself with details. If he does not consider them important, neither will his subordinates. Yet "the devil is in the details." It is hard and monotonous to pay attention to seemingly minor matters. In my work, I probably spend about ninety-nine percent of my time on what others may call petty details. Most managers would rather focus on lofty policy matters. But when the details are ignored, the project fails. No infusion of policy or lofty ideals can then correct the situation.

To maintain proper control one must have simple and direct means to find out what is going on. There are many ways of doing this; all involve constant drudgery. For this reason, those in charge often create "management information systems" designed to extract from the operation the details a busy executive needs to know. Often the process is carried too far. The top official then loses touch with his people and with the work that is actually going on.

I require frequent reports, both oral and written, from many key people in the nuclear program. These include the commanding officers of our nuclear ships, those in charge of our schools and laboratories, and representatives at manufacturers' plants and commercial shipyards. I insist they report the problems they have found directly to me—and in plain English. This provides them unlimited flexibility in subject matter—something that often is not accommodated in highly structured management systems—and a way to communicate their problems and recommendations to me without having them filtered through others. The Defense Department, with its excessive layers of management, suffers because those at the top who make decisions are generally isolated from their subordinates, who have the first-hand knowledge.

To do a job effectively, one must set priorities. Too many people let their "in" basket set the priorities. On any given day, unimportant but interesting trivia pass through an office; one must not permit these to monopolize his time. The human tendency is to while away time with unimportant matters that do not require mental effort or energy. Since they can be easily resolved, they give a false sense of accomplishment. The manager must exert self-discipline to ensure that his energy is focused where it is truly needed.

All work should be checked through an independent and impartial review. In engineering and manufacturing, industry spends large sums on quality control. But the concept of impartial reviews and oversight is important in other areas also. Even the most dedicated individual makes mistakes—and many workers are less than dedicated. I have seen much poor work and sheer nonsense generated in government and in industry because it was not checked properly.

One must create the ability in his staff to generate clear, forceful arguments for opposing viewpoints as well as for their own. Open discussions and disagreements must be encouraged, so that all sides of an issue will be fully explored. Further, important issues should be presented in writing. Nothing so sharpens the thought process as writing down one's arguments. Weaknesses overlooked in oral discussion become painfully obvious on the written page.

When important decisions are not documented, one becomes dependent on individual memory, which is quickly lost as people leave or move to other jobs. In my work, it is important to be able to go back a number of years to determine the facts that were considered in arriving at a decision. This makes it easier to resolve new problems by putting them into proper perspective. It also minimizes the risk of repeating past mistakes. Moreover, if important communications and actions are not documented clearly, one can never be sure they were understood or even executed.

It is a human inclination to hope things will work out, despite evidence or doubt to the contrary. A successful manager must resist this temptation. This is particularly hard if one has invested much time and energy on a project and thus has come to feel possessive about it. Although it is not easy to admit what a person once thought correct now appears to be wrong, one must discipline himself to face the facts objectively and make the necessary changes—regardless of the consequences to himself. The man in charge must personally set the example in this respect. He must be able, in effect, to "kill his own child" if necessary and must require his subordinates to do likewise. I have had to go to Congress and, because of technical problems, recommended terminating a project that had been funded largely on my say-so. It is not a pleasant task, but one must be brutally objective in his work.

No management system can substitute for hard work. A manager who does not work hard or devote extra effort cannot expect his people to do so. He must set the example. The manager may not be the smartest or the most knowledgeable person, but if he dedicates himself to the job and devotes the required effort, his people will follow his lead.

The ideas I have mentioned are not new—previous generations recognized the value of hard work, attention to detail, personal responsibility, and determination. And these, rather than the highly-touted modern management techniques, are still the most important in doing a job. Together they embody a common-sense approach to management, one that cannot be taught by professors of management in a classroom.

I am not against business education. A knowledge of accounting, finance, business law, and the like can be of value in a business environment. What I do believe is harmful is the impression often created by those who teach management that one will be able to manage any job by applying certain management techniques together with some simple academic rules of how to manage people and situations. (COPYRIGHT 1981, H. G. RICKOVER)

All of these principles can observed on any given day on any given Nuclear Submarine. The root of the Nuclear Submarine's track record of operational excellence stems from the work, passion, and philosophy of one man: Admiral Hyman G. Rickover.

It is worth repeating that this book is neither a memoir nor a book of boastful accomplishments by the authors or the Nuclear Submarine community. The book was written with the aid of a criterion that we referenced over and again while writing and editing, "Will this content provide actionable value to a reader who is committed to leading his organization to a standard of operational excellence?"

If you are a leader in an organization that is somewhere in the spectrum from interested to fully committed to raising the standards of your organization's culture and performance, we wish you the best of luck. Further, we would like to remind you that

Operational Excellence is not a destination—it is a journey. Please feel free to contact us at ExtremeOperationalExcellence. com, if you seek assistance in your journey.

Very Respectfully,

Lieutenant Commander
Matthew DiGeronimo
USN, Retired

Commander
Bob Koonce
USN, Retired

CHAPTER 1

OPERATIONAL EXCELLENCE

"Emergency Report. Emergency Report. Fire in the Engine Room. Fire in Engine Room Upper Level! Fire in the Port Switchboards!" bellows the Engine Room Upper Level Watch over the ship's emergency announcement circuit heard by the controlling stations and the Commanding Officer's stateroom.

Before the announcement reaches its last syllable, the ship's Auxiliary Electrician arrives at the scene with two EABs (Emergency Breathing Apparatus) and a fire extinguisher. He dons his breathing mask as he hands the other one to the Engine Room Upper Level Watch.

"Fire in Engine Room Upper Level. Fire in the Port Switchboards. The fire main is pressurized. Firehose 7 is primary. Hose 6 is backup," announces the Chief of the Watch on the 1MC (general announcing circuit).

The General Alarm is then sounded. *Dong. Dong. Dong. Dong. Dong. Dong . . .*

The 1MC announcement and General Alarm is heard by the entire crew.

Every man on the submarine springs into action. There is not a man aboard the ship who does not have a specific action to accomplish in response to a fire. Now is not the time for thoughtful deliberation. Muscle memory developed from extensive training assumes control as each man executes his duties as rapidly as humanly possible. Each man moves throughout the submarine at a pace just short of sprint. Their bodies move up and down ladders with the ease and speed of squirrels scurrying up a tree. Those that were sleeping are fully dressed and interspersed with the crew in a matter of seconds.

The Engineering Officer of the Watch gives the order to secure the power to affected switchboards and Electrical Operator responds rapidly. The Maneuvering Area is a buzz with action. The loss of an electrical switchboard power has an impact on the nuclear reactor and steam plant that must be handled swiftly and precisely correct. Also, as the controlling station in the Engine Room, the Maneuvering Area watchstanders don emergency breathing apparatus then seal and pressurize the space to prevent smoke from entering.

The Engine Room Upper Level Watch and Auxiliary Electrician discharge two extinguishers on the affected switchboard. Before the charge on the second extinguisher is spent, the remaining Engine Room Watchstanders have taken Hoses 6 and 7 off their stowage reels and routed them to the scene.

With Hose 7 routed to the scene and manned with a Nozzleman stationed, the Engineering Watch Supervisor yells, "Pressurize Hose 7!" The Engine Room Lower Level Watch opens the isolation valve and Hose 7 is instantly pressurized and ready for use. With the rapid response of fire extinguishers and the immediate isolation of power to the switchboard, the electrical fire is out, but the flames that escaped the switchboard spread to the bulkhead insulation. Hose 7 attacks the flames while Hose 6 is pressurized, manned and ready for use in Engine

Room Middle Level in anticipation of the fire spreading to level beneath its seat—which it does.

A fire is a submariner's worst nightmare. The enclosed space of a submarine appears large when touring the area that encapsulates the existence of 140 men for months at a time, but even a small fire can shrink that perception in a matter of seconds. Within 60 seconds light smoke is visible in every compartment. In another 60 seconds visibility is slightly reduced in the forward compartment and all but lost in the Engine Room. The hose teams are fighting a fire they can't see through the thick black smoke. One member on each hose team is using a thermal imager that enables him to see the fire through the black smoke. He communicates to the Nozzleman how to direct his efforts (move the stream up or down, left or right, sweep up and down or left and right, hold steady, etc.) through a coded series of taps on his back—voices are drowned out and practically useless.

The breathing masks that the hose teams are using are supplied through a six-foot cord that is manually plugged into the ship's low pressure air system. This puts each man tethered to within six feet of one of the dozens of manifolds scattered about the ship. To move beyond these six feet, a man must take a deep breath, unplug his cord, and then move swiftly to the manifold closest to his destination while holding his breath. Without countless hours of practice, this would be an impossible feat in this smoke-filled compartment. That is precisely why each submariner is forced to practice moving about the ship blindfolded while wearing an EAB and traversing the length of the ship from forward to aft and top to bottom.

The fire continues to spread in the Engine Room. Ensign Jones, the Man in Charge of Hose 7, knows that he must let

Damage Control Central (the Commanding Officer's state-room is utilized as a makeshift command center during casualties) know the status of the fire—the fire has spread to Engine Room Middle Level and continues to spread in Engine Room Upper Level. Unfortunately, he cannot find his phone talker. He unplugs from his EAB so that he can reach an installed phone. However, as he approaches the phone, he cannot locate the air manifold. He begins to panic, his lungs are on fire, and his head grows light. In an act of desperation, he lifts the mask from his face in an effort to obtain a breath of air, but, instead, he fills his lungs with smoke. He is overcome and collapses to the deck.

In the control room, the Officer of the Deck has driven the ship to periscope depth in preparation for ventilating the Engine Room. However, until the fire is out, ventilating the Engine Room would exasperate matters by providing the fire with a fresh supply of oxygen, thereby making the situation much worse. The crew must rely on the hose teams to put the fire out.

The heat from the spreading fire is too intense for the hose team members to tolerate. They are pushed back too far from the fire for hoses to be effective. Just then, two new hose teams donned in full firefighting gear, including self-contained breathing tanks, arrive at the scene to relieve the unprotected fire teams. Their insulated protective gear allows them to move the hoses closer to the seat of the fire. The two hoses stop the spread of the fire. They contain the fire. They fight and fight. They are reaching the limits on their air tanks, but the fire is relenting. Moments later the fire is out.

"The fire is out!" the Nozzleman shouts. The Executive Officer, who reports to the scene of a casualty as the Man in Charge, reports over the ship's emergency announcing circuit:

"Emergency Report. This is the XO. All fires are out." The Chief of the Watch then informs the entire crew over the 1MC: "All fires are out."

The Executive Officer orders the reflash watch to be set by both hoses, as the Officer of the Deck orders ventilating the Engine Room. The Executive Officer then orders the thermal imaging operator to search the Engine Room for injured personnel. As the smoke begins to clear from the Engine Room, the hose teams stare intently at the seat of the fire, knowing full well that ventilating, although necessary to remove the smoke and restore visibility, also creates an oxygen-rich environment capable of causing a reflash fire. There are no signs of reflash.

"Injured man! Injured man in Engine Room Upper Level!" The search for injured personnel yields the discovery of Ensign Jones who was overcome by smoke while attempting to report the spread of the fire in the Engine Room.

The Executive Officer orders the ship's Corpsman to report to the Engine Room and attend to the injured man, although he knows that there is nothing that can be done for Ensign Jones other than to respectfully remove his body from the Engine Room.

Suddenly a clear and fresh voice announces on the 1MC: "Secure from the drill. The fire in Engine Room Upper Level was a drill conducted for training. Restow all damage-control gear and secure from the drill."

After a quick debrief from the Engineer Officer about the crew's performance, the Commanding Officer addresses the crew from the 1MC: "Great effort by everyone. We ran a challenging scenario today. As you know, we expect an electrical fire to be extinguished when power is secured to the affected switchboard. However, today we tested our ability to fight a spreading multilevel fire. We haven't collected all of the comments from the drill team yet, but here are a few big-picture feedback points. We had a man overcome by smoke in the Engine Room who would not have survived. This is, obviously, not acceptable, and we will be looking into the root causes behind that event. On a positive note, we met all of our timing metrics for locating the fire, reporting the fire, securing power, expending a fire extinguisher, having a hose fighting the fire, and the relief hose teams on scene. That is very encouraging. We may need to make those metrics shorter to challenge you more. [Pockets of laughter can be heard throughout the boat]. Again, great effort and expect the drill team's comments later in the day for your review."

The time is 0730. Those that had the midwatch last night are now desperate for sleep. However, every crew member is drenched with sweat. There is only one option—shower first, sleep second. The crew shares four shower stalls. The lines are long but move along. Submariners learn to shower quickly. Not only because there is most always some waiting on you, but submarines make their own water and learn to conserve each drop. A submarine shower is *turn on the shower, get wet, turn off the shower, lather up, turn on the shower to rinse off, and get out.* Anything more is called a "Hollywood shower" and not allowed or accepted aboard a Nuclear Submarine.

Casualty drills are part of life aboard a submarine. Take the most dangerous possibilities and then prepare the crew to fight and overcome scenarios that are even more dangerous. The necessity of this training doesn't hit home until a submariner encounters his first actual casualty. When a real fire does occur on a submarine, the response can only be described as a "thing of beauty"—the fire doesn't stand a chance. However, even the smallest fires can fill the submarine's atmosphere with smoke quickly—very quickly. When a submariner sees this for the first time, a switch goes off in his mind. He "gets it." He understands the purpose of the fire drills at 0200. He understands the purpose of simulating raging fires. The crew's proficiency is truly a matter of life and death.

The damage control gear is cleaned, inspected, and restowed. The fires hoses are drained to the bilges and then restowed. The shower lines diminish. The section that stood the midwatch are asleep. The submarine is back to its normal routine, knowing and accepting, that its routine is always subject to disruption by a casualty drill. However, the work of increasing the crew's proficiency at responding to fires has just begun.

It takes a lot of effort and coordination to run a drill on a Nuclear Submarine. The "drill team" is a group of people who provide simulations and indications to the crew to maximize training value. The fire itself is simulated by a string of red lights. Smoke is simulated by placing vision impairment devices (usually shower caps) over the crew's EAB masks to simulate reduced visibility. Simultaneously, the drill team is vigilantly defending the safety of the crew during the drill. There are no soft surfaces on a submarine, so when the entire crew is engaged in a fire fighting drill with shower caps on their faces, the risk of an injury is high. Additionally, drill team members are taking copious notes about the crew's response. These notes will formulate a list of deficiencies that will be used to provide feedback to the crew. Perfection is unobtainable; even in a drill that would be considered "excellent," there will be a list of deficiencies or areas for improvement. In fact, the sharper and more proficient the crew becomes, the more demanding the drill team's expectations. As a result, the bar never stops moving higher.

Keep in mind that throughout this entire scenario, this 360-foot long, 6900-ton nuclear powered vessel is travelling submerged at depths up to 800 feet below the surface of the ocean with a complex nuclear propulsion plant providing power. At no time can the operational team let down its guard for safe operation of this warship. The ocean is very unforgiving, even, and especially, when running casualty drills for training.

In case you're wondering, Ensign Jones—the one who "died" in the fire drill—spent the next two hours traversing the boat from the XO's stateroom aft to the Engine Room Lower Level in an EAB with a shower cap distorting his vision, practicing moving from EAB manifold to EAB manifold. He'll do the same drill tomorrow, and the next day, and the next day until the Commanding Officer trusts that his ability to do so is the best on the ship. "I'm going to take you from worst to first, Ensign Jones!" said the Commanding Officer as he put his right arm around Ensign Jones shoulders with a fatherly tone. "You are going to be so good at EAB operations, that the next time we have a crew member who struggles with his EAB during a drill, guess who I'm going to assign to upgrade him?"

Although we haven't yet defined *Operational Excellence* for you, we wanted to make it clear from the beginning that the journey towards Operational Excellence isn't an easy or painless one. The journey requires blood, sweat, and tears—usually figuratively.

WHAT IS OPERATIONAL EXCELLENCE?

Operational Excellence. It sounds like something we all want to be a part of, but, what is it? This entire book is about the pursuit of Operational Excellence so we'd like to make it clear from the beginning what we are pursuing. There are many definitions of this elusive descriptor and a lot of variation between those definitions.

Operational Excellence is conventionally and typically defined by its output and sounds something like *"efficient and optimized levels of performance resulting in strong financial and operational performance."* Often, the definition is dressed up with more complex and academic language, but if you dig to the heart, most definitions will boil down to this same output-based definition. We believe that, although this is a fairly accurate definition of the output of Operational Excellence, it falls short of defining Operational Excellence itself. We don't typically define things by their output. We wouldn't define a turbine generator as "electricity" or a human being as "carbon dioxide." Instead, this is our definition of the term that drives each chapter of this book:

Operational Excellence is a culture fully devoted to professional knowledge, brutally honest self-assessment, continuous improvement, and intellectual integrity.

However, Operational Excellence is like pot of gold at the end of the rainbow. It cannot be reached. Functionally, it is a journey, not a destination. Continuous improvement is a central core of Operational Excellence; therefore, a theoretical declaration of "We have achieved Operational Excellence" will have the effect of reducing the level of Operational Excellence of that boastful organization. We recognize there is some word play here, but the concept we are driving at is anything but playful.

The leaders of organizations must recognize that their organizations are composed of imperfect people performing their duties imperfectly. That is true of the worst and the greatest organizations. Losing sight of this fact is one of the singular most damaging things that can happen to an organization that is pursuing Operational Excellence because it will, due to the frailties of human nature, reduce the effort that the leaders of the organization take to identify, expose, and address the weaknesses and imperfections of the organization.

We must admit that our definition creates an awkward semantic situation. Consider an organization that is fully committed to the journey towards Operational Excellence. According to our definition, we can reference only its effort to achieve Operational Excellence. The most obvious example of this is the Nuclear Submarine force. We refer to its history of Operational Excellence, when we really mean its history of striving for Operational Excellence. However, we have chosen to use this language to acknowledge that there is a spectrum along which organizations fall in their striving for Operational Excellence. We consider this a shorthand notation, but it is critical to make this distinction.

We've alluded to the outputs of Operational Excellence and provided the definition of Operational Excellence. What about the inputs? The inputs required for Operational Excellence are: 1) leaders and workers who put truth before ego and perception, 2) processes that hold the organization responsible to unwavering standards, and 3) integrity.

In most literature on the subject, the outputs of Operational Excellence receive the lion's share of attention when, in fact, they are simply a logical outgrowth of the inputs and the Operational Excellence journey. This is important to acknowledge because often times these outputs can be achieved without the necessary backing of the cultural elements of Operational Excellence. For example, if an organization's major competitor goes out of business, we would expect to see short-term results that mirror the outputs that we have defined for Operational Excellence. Another, more specific, example is one of a pharmaceutical company that launches a one-of-a-kind, remarkably effective drug. The result of this launch is likely to provide the organization's leadership with the perception of Operational Excellence.

Figure 1.1

Figure 1.2

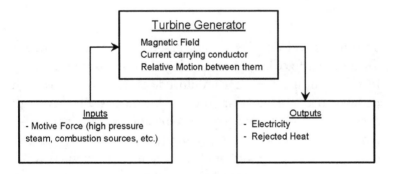

Figure 1.1 provides a graphical representation of our definition of Operational Excellence as well as its inputs and outputs. Figure 1.2 provides an analogous structure of inputs and outputs using a turbine generator, which shows you visually how you could easily obtain electricity and rejected heat without

actually having a turbine generator. There are many ways to obtain electricity and rejected heat. Further, you may have a motive force such as high pressure steam, but that in no way implies you have a turbine generator. It only implies that the turbine generator requires that motive force to function.

Consider an organization that has the following:

- Leaders who put truth over ego
- Processes that enforce standards
- Integrity
- Efficient and optimized levels of performance
- Financial success

Might it be reasonable to conclude that this organization has achieved Operational Excellence? Our answer: not necessarily. The same way that a system that contains a motive force, electricity, and rejected heat does not necessarily have a turbine generator.

The inputs and outputs are not descriptive enough to define Operational Excellence because they do not ensure the stability of the future. There are a multitude of ways to achieve high performance with financial success, but to **sustain** this result over an extended period requires Operational Excellence. This requires the leaders of an organization to look beyond this quarter's earnings report or their current state of success because without a culture of Operational Excellence these results, even with the necessary inputs, will be fleeting at best and luck-based at worst.

WHY BUILD A MODEL OF OPERATIONAL EXCELLENCE FROM NUCLEAR SUBMARINES?

There are thousands of organizations on the planet. Many of them have strong cultures and track records of success that are worthy of credit and recognition. We have no delusions that the Nuclear Submarine community's culture holds the only key to Operational Excellence. However, there are many unique characteristics about the Nuclear Submarine culture that makes its pursuit of Operational Excellence unique, fascinating, and an effective to model to mirror. Those characteristics include:

1. Human nature at its rawest. There is nowhere to hide on a Nuclear Submarine. There are no vacations or sick days while you are at sea. There are no holidays or weekends. The work is 24 hours a day until the ship returns to the pier. During that period, which can range from days to months, there is little if any contact with the outside world. Even in today's world of connectivity, when a submarine goes into stealth mode there is a shutdown of all communications. One of the consequences of living in this environment is that everyone's true self shines through. We have all seen each other moments after waking up without the benefit of a shower or a cup of coffee. Working towards a goal with someone after three months underway is the equivalent of working with someone for years, if not decades, in a normal work environment. The submarine lifestyle is a petri dish for human interaction, conflict resolution, and mission accomplishment.

2. The culture is the straw that stirs the drink. In other organizations, a leader may hold the position for 10 or more years. The same inner circle of people may have been running the organization from their various

positions for decades. In these instances, individual talents and leadership styles can drive an organization's operations. On Nuclear Submarines, there is no one who spends more than five years aboard the ship, and most leaders spend less than three years aboard. This creates a revolving door of operators, technicians, mechanics, and leaders. However, the show must go on with whatever group of people is assigned to the submarine at any given time. Many organizations could not survive this type of turnover. The Nuclear Submarine Force excels in spite of this turnover because it has been forced to make provisions for it.

3. Less variables than in most organizations. The Nuclear Submarine community's performance does not sway with interest rates, the stock market, the real estate market, or the price of oil. It is immune from trends and bubbles and has no competition other than itself. This is not to suggest there is no change to deal with. Submarines are routinely implementing new technology and tasked with assignments of increased complexity and risk. However, the minimal external factors allow the community to measure its performance in an environment that is virtually devoid of noise-producing variables. Nuclear Submarines do not make sweeping organizational changes because profits or stock price are down for two consecutive quarters due to deflated oil prices. While this does eliminate real-world variables from their operations, it also results in a sustained focus on cultural elements and striving for Operational Excellence.

4. Sneak preview on generational differences. The average age of a crew member on a Nuclear Submarine is 25. Due to operational necessity, crew members are given

increased levels of responsibility compared to their civilian counterparts. As a result, the Nuclear Submarine community is able to observe the youngest adult generation develop and perform before the rest of the world expects them to excel.

5. Zero defect standard. The Nuclear Submarine community is held to a zero defect standard on two fronts. The first is nuclear safety. The public will not be exposed to increased radiation or contamination levels from Navy nuclear reactors. Period. Secondly, our submarines will surface the exact same number of times that they submerge. Should any singular Nuclear Submarine fail on these fronts, the safety of the public and the lives of the crew, respectively, would be in jeopardy. Neither of these outcomes is acceptable, and, therefore, the Nuclear Submarine community must maintain the highest standards of performance at all times.

6. Nuclear Submarines are fascinating. Let's face it, the nuts and bolts of cultural and organizational Operational Excellence can sometimes force us to consider elements that are a bit less exciting than a day at a water park. Discussing these elements as they apply to the operation of the United States Military's most advanced war machine whose culture and missions are cloaked in a veil of secrecy can spice up the conversation a bit.

7. It's what we know best and are passionate about. Since retiring from the Navy, we have, collectively, racked up experience in small entrepreneurial endeavors; politics; consulting for large corporations, small businesses, and individuals; mergers and acquisitions; commercial real

estate; corporate training; utility operations; design and building of power plants; talk radio; and business development. We have each observed that some elements of culture that make the United States Submarine Force successful could be of great benefit to civilian organizations.

WHY US?

Okay so the Nuclear submarine community represents the pursuit of Operational Excellence from several unique dimensions. But why us? Why should we be the ones translating and sharing this pursuit. The answer is simple: no good reason other than our desire to share something of value with the world. This is NOT a memoir or our professional swansong. In fact, it is tribute to our mentors and shipmates who helped teach us the lessons this book shares. Further, there is virtually no guidance in this book that we aren't or weren't guilty of needing ourselves. Sometimes the guidance we share comes directly from our own leadership weaknesses that we experienced in our careers. We are students of leadership and Operational Excellence first and foremost. This book attempts to capture some of the larger lessons that we would have appreciated receiving throughout our Naval careers—many of which were learned only through hindsight and introspection of our time aboard Nuclear Submarines.

ROADMAP

Spending months at a time with 130 people executing missions vital to national security while operating a nuclear power plant and completely isolated from the rest of the world is very similar to … nothing. There is nothing to compare the lifestyle to. But the glue that holds the crew together through thick and thin is the deep-rooted culture of Operational Excellence. This culture is not only necessary to our survival, the culture is also what underpins the success that Nuclear Submarines have enjoyed for decades. You may not have heard much about that success; this further amplifies the point. It is called the "Silent Service" for a reason. Nuclear Submarines were built for and are utilized to execute secret operations. If stories of Nuclear Submarine operations were gracing the front page of our newspapers, that would indicate, most likely, a problem. Very few people are privy to the details of Nuclear Submarine operations. In fact, it is not uncommon for only a handful of the submarine crew to be knowledgeable of the details of a mission.

We are certainly not going to be revealing any of the information to which we have sworn an oath of secrecy. What we will be sharing are the specifics of the culture that enables the submarine crews to be as successful as they have been. We are certain that tales of clandestine submarine operations would have you on the edge of your seat, but we can't tell those stories. What we can share are the details of the culture of Operational Excellence that can help you and your organization as you strive towards Operational Excellence. That is our mission in writing this book.

The elements that comprise this culture of Operational Excellence can be described using a metaphor of a building.

The foundation is *knowledge* and *learning*. The submarine environment is extraordinarily technical. Not only must the nuclear power plant be operated meticulously correctly but so must every other system on the submarine. If the crew failed to have the requisite technical knowledge, our metaphorical building would collapse. However, knowledge is just a part of the foundation. The pursuit of more knowledge—learning—is the stronger, but subtler, component of the foundation. These concepts will be elaborated upon in Chapter 2.

Standards are the door. Every organization must have principles that each member of the organization must subscribe to. Consider them minimum entry fees. By themselves they do not assure success, but without them, you can't even get inside. Of course, doors work in both directions. If a member of an organization cannot embrace the organization's standards, the door works well as an exit. We will be focusing our discussion of standards on the Nuclear Submarine community's standard of procedural compliance. In Chapter 3, we will discuss what makes a successful standard and why some organizational standards fall on deaf ears.

Our metaphorical building has two weight bearing beams: *questioning attitude* and *watchteam backup*. The former addresses a specific way of thinking and the latter a specific way of acting. Both orbit around the notion that in order to achieve continuous improvement and then leverage that continuous improvement, organizations must create an environment that supports its members in questioning everything—from the most basic to the most nuanced principles of the organization. However, both questioning attitudes and watchteam backup can run amok if not cultivated and led properly. We will be discussing these topics in Chapter 4 and 5.

Integrity is the building's roof. It is the element of the building that is seen the least but when inclement weather arrives, it protects us. Integrity is related to but not nearly as straightforward as "honesty." Integrity in the context that we present it, is "doing the right thing, even when there is no one watching." Without integrity, a Nuclear Submarine would be exposed to the elements, which are harsh and unyielding. We elaborate on this concept in Chapter 6.

After reading through Chapter 6, you will identify a theme. That theme is identifying and tackling mistakes head on with brutal *honesty*. The culture relies upon imperfections to be identified so that they can be evaluated, which, invariably, produces an opportunity to improve the Nuclear Submarine's processes. This cycle ensures that continuous improvement is a way of life not just an expression on a break room poster. The process that epitomizes the evaluation of an imperfection or mishap is the Nuclear Submarine's "critique" process—some organizations call this an incident investigation process, but it is more than that. It a powerful tool for identifying the specifics of what happened in an efficient manner, and most importantly assigning the appropriate corrective actions to prevent recurrence of the mishap. The secret to success of this process is when the people involved in the incident and the ship's operational leadership come together to resolve why this happened and how to prevent it from happening again. This is the culture-impacting opportunity that is missing in many organizations. This critique process is discussed in detail in Chapter 7.

In a critique, the leaders must appropriately identify the root cause(s) to the problems that led to the mishap. The corrective actions are designed to address these root causes. Therefore, if the root cause is not identified correctly, it is highly likely that the corrective actions will be ineffective. Therefore, the chance of recurrence has remained unchanged despite all of the time and resources invested in the critique process. Root cause analysis is the most challenging element of the critique process; therefore, we devote an entire chapter—chapter 8—to the process of root cause analysis.

The implementation of the five components of a culture of Operational Excellence requires strong leadership. However, not all leadership is created equal. In fact, there is no one-size-fits-all leadership style or philosophy. An organization's performance has a natural tendency to sway in a pendulum-like pattern. When a leader enters an organization, the most important assessment she must make is where is the pendulum and in what direction and with what speed is it moving. The second assessment is what do the existing leaders think about their pendulum. If the new leader assesses that the organization is mediocre but getting worse slowly and the existing leaders assess that the organization is above average and getting better quickly, this mismatch will create a unique leadership challenge that must be acknowledged and addressed with the appropriate leadership skills. We explore this concept in chapter 9.

Chapter 10 is a summary. Not many of us take notes when reading a book, but most of us wish we had from time to time. Should you ever need to refer back to this book in the future, Chapter 10 will serve as your guide to the highlights.

We sincerely hope that you enjoy reading this book. More importantly, we hope that you are able to bring some new lessons and tools on Operational Excellence to your organization. If we can help you in your journey towards Operational Excellence, please connect with us at ExtremeOperationalExcellence.com

CHAPTER 2

KNOWLEDGE AND LEARNING

The concept "knowledge is power" is referenced so frequently that it has become a cliché to call this concept cliché. However, it bears repeating because there remains an unclaimed competitive advantage at its core available for the taking for many organizations. Select 100 random articles from the LinkedIn, Harvard Business Review, Forbes, etc., and it is unlikely that many will be about improving the baseline level of knowledge of an organization. In fact, most articles about business and organization culture implicitly assume that the members of the organization have the requisite professional knowledge needed to make the organization successful. The emphasis is usually about managing and leading the people with this knowledge in an effective fashion and direction. There is no doubt that managing and leading strategies are critical to the success of an organization. However, at the end of the day, if one of two competing financial planning companies is composed of people who have more knowledge than the other company about the financial planning basics, industry trends, and investment tools, that company will likely outperform the other, in time, with all other variables assumed equal.

Knowledge is the foundation of operational excellence. Period. The other aspects that we will address are not sustainable without a base of knowledge combined with the hunger for more. Organizations compete. They compete for people's money, time, or influence. Even the most charitable organizations such as the Red Cross, Salvation Army or United Way must compete. The odds of sustained success increase when there is a sound foundation in the knowledge base of the organization and its members. Without this knowledge and the hunger for more (learning), an organization will not be able to withstand the stresses of competition, a complex marketplace, rapidly advancing technology, and communication mediums that are growing faster and more sophisticated by the minute. Therefore, before launching into a 360-degree evaluation about the cultural components of your organization, take a hard a look at the basic "blocking and tackling" of your organization first—its knowledge base and its desire to expand that base.

The Nuclear Navy has embraced and heightened this concept to a degree that is challenging to describe and is even more challenging to survive. The Nuclear Submarine training pipeline is the most academically strenuous training pipeline in the United States Armed Forces—without question. This training pipeline represents the Nuclear Submarine community's willingness to invest the necessary time, money, and resources into developing operators and supervisors. As a result, graduates of this training pipeline have a knowledge base that far exceeds what is required for these individuals to contribute on day one aboard a Nuclear Submarine or Aircraft Carrier but prepares them to integrate into a culture of operational excellence immediately. Their knowledge is the foundation that supports the other components that will be outlined later in this book: questioning attitude, watch team backup, procedural compliance, and integrity.

Each organization manages its training pipeline uniquely. Some have none, while others have exhaustive programs. The next section addresses the specifics of the Nuclear Submarine training pipeline. We present this information to provide context to the submarine anecdotes that you will read in this book. Not every organization can, or should, provide a training pipeline as robust as the Nuclear Submarine community's training pipeline. It is not our intent to imply otherwise. However, we anticipate that many readers have no knowledge of what type of training is required to become a crew member aboard a Nuclear Submarine; therefore, we thought we would take some time at this early stage in the book to detail the training regime required of submariners.

THE NUCLEAR SUBMARINE TRAINING PIPELINE

The requirements to be considered for inclusion in the all-volunteer service of the Nuclear Submarine community are robust and unwavering. Enlisted members must register test scores on the Armed Services Vocational Aptitude Battery (ASVAB) higher than any other Navy subspecialty. Academically, the easiest way to describe the requirements for being accepted into the enlisted nuclear field is that each applicant would also meet admission requirements for the strongest college engineering programs in the nation.

After the successful completion of basic training (boot camp), candidates must complete an "A" school, which is a non-nuclear technical school, as an electronics technician, electrician, or mechanic where they learn the basics of those respective trades. After successful completion of "A" school (between 13 – 26 months), graduates begin the 26-week Navy Nuclear Power School program.

These schools are as academically demanding as any other academic program of which we are aware. Both "A" school and Nuclear Power School consist of eight hours of college-level instruction a day, Monday through Friday, followed by two

– five hours of homework and studying. All of the information is cumulative. If a student shuts down mentally for more than a day . . . adios, amigo. Stand at the train station with your bags packed, orders to a conventional surface ship in your hand, and wave goodbye to your future submarine friends. Literally.

The intensity of the year-long program is a shared experience among graduates that words fall short of describing. When an 18-year-old mechanic graduates Nuclear Power School, he understands the specifics of nuclear fission in a pressurized water nuclear reactor exhaustively. He can discuss the minute details of how fission works, how much energy it creates, how it responds to changes in operational parameters, the metallurgy of the plant, and even the specifics and reasons for the chemistry control of primary coolant. A reasonable person could conclude that it is unnecessary to demand this much knowledge of the operators. However, this standard of education has a sixty-year heritage and has produced an impeccable record of success.

Officers hoping to join the Nuclear Submarine community attend Officer Nuclear Power School after graduating college. Their school is also 26-weeks long with the same format of instruction, studying, and homework. Approximately 40 percent of the officers are engineering majors from the Naval Academy. The other 60 percent consist mostly of engineering majors from the nation's strongest engineering programs. Interestingly, an engineering degree is not a prerequisite to the program—extraordinarily bright and talented biology, history, and English majors are known to constitute a small percentage of each officer class. To be accepted into the program, the student must pass a gauntlet of verbal interviews with nuclear engineers in Washington D.C. known as Naval Reactors. There is no requirement for the officers to have any background in nuclear engineering; the interviews test for strong understanding of advanced math and scientific principles and the ability to communicate these principles verbally. If the candidates pass these interviews, they are then interviewed by the Four-Star Admiral that runs Naval Reactors. These interviews are the legacy of Admiral Hyman G. Rickover—the father of the Nuclear Navy—who was notorious for rattling the cages of even the most highly qualified officer candidates. One famous Nuclear Submarine Officer, President Jimmy Carter, recounts his interview with the Admiral:

I had applied for the nuclear submarine program, and Admiral Rickover was interviewing me for the job. It was the first time I met Admiral Rickover, and we sat in a large room by ourselves for more than two hours, and he let me choose any subjects I wished to discuss. Very carefully, I chose those about which I knew most at the time—current events, seamanship, music, literature, naval tactics, electronics, gunnery—and he began to ask me a series of questions of increasing difficulty. In each instance, he soon proved that I knew relatively little about the subject I had chosen. He always looked right into my eyes, and he never smiled. I was saturated with cold sweat. Finally he asked a question and I thought I could redeem myself. He said, 'How did you stand in your class at the Naval Academy?' Since I had completed my sophomore year at Georgia Tech before entering Annapolis as a plebe, I had done very well, and I swelled my chest with pride and answered, 'Sir, I stood fifty-ninth in a class of 820!' I sat back to wait for the congratulations which never came. Instead, the question: "Did you do your best?' I started to say, 'Yes, sir,' but I remembered who this was and recalled several of the many times at the Academy when I could have learned more about our allies, our enemies, weapons, strategy, and so forth. I was just human. I finally gulped and said, 'No, sir, I didn't always do my best.' He looked at me for a long time, and then turned his chair around to end the interview. He asked one final question, which I have never been able to forget—or to answer. He said, 'Why not?' I sat there for a while, shaken, and then slowly left the room.[1]

After the successful completion of Nuclear Power School, enlisted and officer students must then complete a six-month Nuclear Power Prototype training program. Prototype training is where theory meets practice. The prototypes are land-based, fully operational nuclear power plants. Students learn to operate the systems that are representative of the systems on board Nuclear Submarines. For many students, prototype training is their first exposure to a learning model that emphasizes practice over theory. It is not uncommon to see the brainiacs that aced Nuclear Power School, struggle at prototype or the students who barely passed Nuclear Power School to excel in prototype. When you are graded by what you physically achieve in real-time instead of what you can answer on a written examination, you discover the two skillsets often do not go hand in hand. After prototype, students report to their first submarine, except for the officers who make a quick three-month stop at Navy Submarine School where they learn the basics of the non-nuclear systems and of submarine warfare tactics.

The academic demands don't stop after Nuclear Power School. They never stop. The second-highest ranking Admiral in the United States Navy personally reviews the training program records for each Nuclear Submarine and Aircraft Carrier. The program requires up to three hours of training per week and three tests per month for each nuclear-trained operator. This requirement is not malleable. Consider a Nuclear Submarine that is on a mission vital to the nation's national security. One tactical mistake could, at any time, result in the loss of life or

---

1   Carter, J. (1996). *Why not the Best? The first 50 years.* AR: University of Arkansas Press.

an international event covered on CNN that night. During these times, which can last for weeks or months, nuclear training continues—on schedule—every week. The tests continue. Upgrade programs for test failures continue. The level of knowledge bar for every nuclear-trained person on a submarine is lofty and slopes steeply.

This robust training pipeline is a necessity for the Nuclear Submarine community to maintain its track record of operational excellence. Many organizations have similar training programs although few, if any, have ones that are more exhaustive and intense. It is not our intent to share that structure as one that should be duplicated by civilian organizations, but rather to provide context for the remainder of the book.

KNOWLEDGE IS NOT ENOUGH

Knowledge is not a fine wine; it does not improve with age by sitting idly; in fact, it decays. How many of your college classes' final exams could you pass today? High school? Our brains are efficient machines, and they don't act kindly to clutter. Some models of human memory suggest that our brains create memories the way Da Vinci created masterful statues. Remove the parts that don't belong, and what is left is the masterpiece. If we have knowledge that is not used, the brain files it away in a closet where it becomes increasingly difficult to find.

As we alluded to earlier, much of what is taught to Nuclear Submariners during their training pipeline is above and beyond what is required of them to perform in their day-to-day duties. Therefore, an effective training program and learning culture have two objectives: The first is to prevent knowledge decay. The second is to push the bounds of each member's learning abilities. There is no one on a submarine (except the Commanding Officer) who is not actively working on a qualification to prepare him for the next and higher position of responsibility. But even the Commanding Officer does not escape the necessity to continue to learn. Before being allowed to take Command, the Prospective Commanding Officer must pass a series of oral and written examinations on the specific nuclear propulsion plant of his submarine at that same Naval Reactors Department where he first was accepted into this training pipeline 15 years earlier. While in command, he is expected to train his crew at least quarterly so he must maintain his level of knowledge.

The point is that "learning" is really the element that more adequately describes the foundation upon which the culture of operational excellence is built. Knowledge is the fruit of learning; this is not a tricky concept, and most organizations are designed to provide its members opportunities to learn and grow through both formal and informal training programs. However, many of these training programs fail to meet their intended objectives because they are not supported by a culture of learning that emphasizes their importance, and that strengthens the roots of the knowledge gained through the training program.

The parochial business adage "grow or die" is being replaced by "learn or die." Developing a learning culture is arguably the most rewarding, productive, and sophisticated leadership accomplishment of our modern world. The Nuclear Submarine community has realized this since its inception. The learning culture is the foundation to the Nuclear Submarine's operational excellence model—without it, the other elements would be fleeting, at best, and non-existent at worst.

As in most organizations, the leadership challenges aboard Nuclear Submarines are vast, complex, and unique. The particular challenges faced vary from moment to moment. As the Navy Seals often say, "The only easy day was yesterday." Leaders who are willing to stand up to the task of creating a culture of operational excellence must embrace the harsh reality that yesterday's challenges are different than tomorrow's. When Albert Einstein was converging on his historic Theory of Relativity, he was blessed with the benefit of working with a finite number of variables that behaved well while he slept. Leaders do not have that luxury. The variables of an organization are gremlins that move, morph, and multiply while we sleep. Yesterday's solutions are tomorrow's damaging initiatives. This dynamic has no "one size fits all" solution. There is no one person, program, or philosophy that can handle the challenges of leading an organization to operational excellence through the vicissitudes of its lifecycle. The solution is leadership that can invigorate the minds of a team and then leverage those ideas to create a collective conscientiousness that is strong enough to prevail against the headwinds of uncertainty and change.

Establishing a foundation of a learning culture is not a destination—it is a journey. This foundation was established decades ago in the Nuclear Submarine community and now runs smoothly, but if you find yourself facing the opportunity to create a new learning culture you may feel like you are running at maximum speed on a treadmill reaching for a carrot just beyond your grasp.

Many organizations look beyond or oversimplify the concept of creating a learning culture. Do the following responses sound familiar?

- We already have a training program.
- We participate in the Continuing Education Program of our industry.

- We have an entire department devoted to training and professional development.
- We encourage our personnel to cross-train.
- We pay for our employees to continue their formal education.

Training programs can yield great results when effectively led. However, how many times have you sat through a one-hour training session that had PowerPoint slides with massive amounts of information, but then you left without learning anything? Are these programs "nice-to-haves" with marginal impact on the organization, or are they fundamentally improving the culture of operational excellence in the organization?

LEARNING CULTURE

A learning culture is defined by what it knows, not what it is. A learning culture:

- Knows that truths are more important than egos.
- Knows that truths are discovered only through brutally honest and deep introspection.
- Knows that truths are often uncomfortable and inconvenient.
- Knows that senior personnel are not right because they are senior.
- Knows that knowledge is inferior to knowing how to learn.
- Knows that effective communication is a master skill, not an entry level one.
- Knows that people want to be challenged but fear being wrong.

- Knows that people have been conditioned to hide their shortcomings.
- Knows that cross-training across specialties is invaluable.
- Knows how little it knows.

The last bullet, "knows how little it knows," is the heart and soul of the Nuclear Submarine's learning culture. If we could put aside the healthy, good-natured, and long-standing rivalry among the different communities of the United States Military, there would be little, if any, debate that the Nuclear Submarine community maintains the highest standards of Operational Excellence. Ironically, there would also be little debate that the Nuclear Submarine community strives to improve its standards more than any other military community.

Variations of Socrates' words of, "All I know is that I know nothing." have been echoed by the most cultured, educated, and intelligent people of all ages.

"He who knows best knows how little he knows." Thomas Jefferson

"The more I learn, the more I realize how much I don't know." Albert Einstein

"How little we know of what there is to know." Ernest Hemingway

The foundation of the Nuclear Submarine's culture is embracing this humble approach to the evaluation of our own knowledge. This embrace is reflected in the establishment and protection of an effective and supportive learning culture where striving for the truth always outweighs the protection of anyone's ego—including, and especially, the senior personnel. On a Nuclear Submarine, this learning culture has been established by generations of submarine crews that were able and willing to put the performance of the crew members and their commitment to continued operational improvement ahead of their egos. This learning culture is the foundation of our operational excellence structure. Without this solid foundation, the other elements that we will be discussing will be transient at best and artificial at worst.

On the other hand, experienced supervisors often become pre-emptively defensive about existing processes. This defensiveness obstructs open and honest dialogue. Some typical knee-jerk responses include "we've always done it that way," "never had a problem before," or "a previous management team wanted it that way." Often these obstructions are a reflection of a culture that is intolerant to the most challenging three words for experienced supervisors to utter -- "I don't know. " Strong organizations seek the truth, but accept and respect "I don't know" as a reflection of members' credibility and intellectual integrity.

In truth, "I don't know" is an expression that adds value, credibility, and opportunity to virtually every aspect of an organization's operations. Unfortunately, it is also an expression that our cultures have attempted to squash, erase, and discourage. Professionals have been conditioned to say almost anything before those three words. A learning culture revitalizes the acceptance of those three words: "I don't know."

There are three major ways of saying "I don't know" without saying those words. They are:

1. **The "Don't get out of bed" answer.** There are always reasons not to do something, including getting out of bed in the morning. The answer emphasizes obstacles and ends with an

implied, but noncommittal, recommendation for taking no action.

Q: Would a geothermal project be a viable energy source for this new development?

A: Hmm. Geothermal has complicated components that have a history of failure. Plus, we haven't performed a heat profile study on that plot of land. They are expensive and long; performing one might delay our engineering design.

2. **The "Let me tell you a story" answer.** People who have been around the block long enough have a story for everything. Partially because they have a lot of experience from which to draw and partially because they want an excuse for not trying something new. Again, the answer ends with an implied, but noncommittal, recommendation for taking no action.

Q: Would a geothermal project be a viable energy source for this new development?

A: Hmm. Did you hear about the building in Minneapolis that used geothermal? When the company announced its building design, they got awesome press. That was three years ago-ish. I know the facilities manager in that building and ran into him at a conference last year. He says the entire system is nothing but a headache. Another buddy of mine runs a geothermal facility in California, and he says the same thing.

3. **The "I've been doing this for x years" answer.** When you hear "I've been doing this for 30 years," listen up because you are likely about to hear to a substitute for "I don't know." Usually this answer provides a conservative recommendation that may not or (usually) doesn't respond to the question.

Q: Would a geothermal project be a viable energy source for this new development?

A: Boss man, I've been doing this job for 30 years. I've seen developers try all sorts of sexy new energy technologies. Nothing beats good old-fashioned coal-fired steam boilers.

Note that in each of the three answers:

- The question wasn't answered.
- The answers demonstrated no signs of weakness or uncertainty.
- The person who asked the question has gained to no useful knowledge.
- No one is leaving the conversation with an action plan, such as "I'll do some research and get back to you tomorrow."

Intentionally or not, these answer types are designed to make the question go away. The question is threatening. The question threatens a person's ego which many of us will defend before and above the truth. Egos need these types of questions to be nipped in the bud because they are a perceived threat—threatening to "expose" their ignorance.

KNOWLEDGE AND INTELLECTUAL INTEGRITY

The lack of intellectual integrity in an organization can become so prominent that the leaders grow numb to it. For example, the newest sales associate asks one of the senior salespersons a question: "Tim, I read an article last night that said the new flux capacitors have a new technology that will antiquate our current product line. What do you think about that?"

Tim has no idea what article this kid is talking about, and he hasn't made an effort to follow the technology advancements of the product line in years. He has grown comfortable with selling whatever the company puts in front of him.

How many of us wouldn't be surprised to hear Tim respond: "Kid, there have been rumors about a new generation of flux capacitors for years. It's all bull. Our flux capacitors have been the market leader for a decade because their technology is proven. Even if new technology were to emerge, it would take years to compete with our track record of reliability."

Compare this response to: "I haven't heard of any new technological advances in the flux capacitor in years. Send me a link to the article so I can read it and then we can discuss it."

Tim's response may be right, but it originates from an intellectually dishonest source. Tim is just regurgitating an industry standard response to the rumor of a new mousetrap. This standard response didn't pop out of thin air; it is based on the patterns of the past. But again, it is intellectually dishonest and devoid of thought.

Take a moment to play out the second- and third-order consequences of the differences in these two responses. The gap between the two is significant and, if left unchecked, grows larger every year. Either you or your competitor is identifying this cultural weakness and is taking action to reverse it. What impact would you expect these measures to have on the companies' market shares down the road?

Why does the lack of intellectual remain unchecked so often and in so many places? Because leaders tolerate this intellectual laziness regarding the expected base of knowledge or, worse, exhibit it themselves.

In many cases, the leaders of organizations have enabled an environment where intellectual integrity degrades beyond the laziness illustrated in the previous example. Intellectual integrity can become dangerous—operationally, financially, and literally. When people grow comfortable, consciously or not, hiding weaknesses about their level of knowledge, bad things happen. When people may overstate their confidence level in their degree of knowledge, this is effectively a bet against the house. It is a roll of the dice that less than honest proclamations of knowledge won't blow up in the face of the person who is listening and, likely, believing the information.

It is imperative that we develop an environment where people are committed to putting the interests and well-being of the organization above their egos. To achieve this, the leaders of the organization must create an environment that embraces accurate reports even if they contain qualifiers that are less melodic than half-truths. Every manager wants to hear, "This is true, and I am sure of it." Unfortunately, when that "want" becomes a "demand," reports that should sound like, "This is what I know, and this is what I think I know; let me get back to you" will become "This is true, and I am sure of it."

It is strenuous, at best, and impossible, at worst, to foster an organizational culture of integrity if the members of the organization are not willing to say, "I don't know, and I need help." A person must know what it feels like to actually "know" something for her to adequately identify when she doesn't—especially if the organization relies on that person to seek assistance for the instances when she doesn't. In many organizations, the level-of-knowledge bar is so low that the probability of someone making a costly mistake that could have been avoided by getting assistance is high. As a leader, this type of error is one of the most difficult pills to swallow, because you know that the organization has the capability of handling the situation but because an individual doesn't have the integrity to admit a lack of knowledge, the entire team suffers the consequences.

WHEN AN ORGANIZATION STOPS LEARNING

"Tom" was just hired as the new General Manager working at a coal and natural gas-fired power plant in the Midwest. The power plant had been operating for nearly a century, but its current operation was not profitable, and Tom was hired to change that. You can imagine the variety of cultural challenges that accompany turning around the operations of a century-old power plant with a staff with an average tenure of two decades.

How do you eat an elephant? One bite at a time.

Tom was evaluating every part of every component in every system in the plant like a detective bent at the waist with his magnifying glass. During this evaluation, he asked questions— a lot of questions. Why this? Why that? He committed to learning as much as he could, as often as he could, wherever he was. His most powerful tool to complement the financial reports, technical drawings, and plant diagrams was asking the questions, often to the annoyance of his staff.

Many of the responses were (get ready to cringe): "That's the way we have always done it." (If you hear this uttered in your organization, you have work to do.) This expression means, "I don't know." It is okay not to know, but the "it has always been done that way" expression, even though it may be true, is never an acceptable answer to the question of "Why?"

There was a particular process that caught Tom's eye. He noticed that the five-story boilers have a continuous process called a surface blowdown. To prevent chemicals or contaminants from leaving the boilers and damaging downstream components, a steady "suck" (vacuum) is applied on the top of the water level where contaminants tend to collect like muck on a pond. The steam that is pulled out through this process is a "loss" in the plant because it is not leaving the boiler to do its job—make electricity. Therefore, if the amount of steam sucked off the top of the boiler was lowered, the plant would operate more efficiently (power plant code for "make more money").

The magnitude of this surface blowdown, measured in percentage of total steam flow, was adjusted by the control room operator and averaged about 15 percent. To fully understand why the blowdown averaged 15 percent (why not 14, or 14.5, or 14.75, or 10?), Tom sought out the expertise of the Plant Manager. The Plant Manager, "Vincent," was tremendously knowledgeable but also displayed signs of defensiveness when asked questions about the plant. His reactions were understandable. Vincent was celebrating his 25th anniversary at the plant the first week Tom arrived.

Tom asked, "Vincent, can you explain to me how we determine our surface blowdown percentage?"

Vincent's reply meandered, and he may have eventually gotten to the answer, but through a path that was unnecessarily complicated: "Sure. It's actually not controlled as a percentage. The operator's display indicates a conductivity level, which is an approximation from the sensor installed in the boiler feedwater line. We replaced the sensor a few years ago, but there is a backup that we sometimes use. Then the Shift Supervisor manually samples the chemistry of the boiler water, but that is difficult to do sometimes because he's the only guy available to perform the sample since we cut the auxiliary operators. So the conductivity is an approximation, but we reprogrammed the

PLC [Programmable Logic Controller] of the system, so the display shows a conservative number. Some of the blowdown heat is captured because we preheat the feedwater . . . "

Tom had to stop him. "Vincent. Timeout. Let's back up about a million feet. Before I could even begin to understand what you just said, I need you to describe the process plainly and then work from there. Let's start with: Why do we perform a surface blowdown at all? Please talk to me like I'm a sophomore in high school who is starving to learn because that is all I am trying to do — learn."

Vincent's reply was coupled with a thinly veiled eye roll. "We suck steam off the top of each boiler, because sediment . . .excuse me . . . yucky stuff settles there, like the muck on top of a pond. We don't want that stuff to leave the boiler so we suck it off."

"Okay, great. I'm with you now," Tom replied and then continued, "What "stuff" is being sucked off the boiler beside steam? What's in the 'yucky' stuff?

Vincent was growing impatient. "Conductivity. That's what I was trying to tell you before. Our operators have to estimate the amount sometimes. Actually, the blowdown valve leaks, so the measurement isn't exactly correct."

"Vincent, what is conductivity?" Tom chuckled and continued, "I am your student. Teach me."

Vincent stared at Tom blankly. After a few painfully awkward seconds, Vincent laughed and said, "Tom, I'm not a chemist. You want to know what conductivity is—Google it. I am just telling you that we base our blowdown rates on it."

Tom knew where he was taking this conversation and intended to take Vincent with him, but he feared that he went too far too fast. He took a few steps back.

"Vincent, I can sense your frustration, and I understand. Please know that I am not trying to tear apart your knowledge or even the existing process. However, I think it's reasonable for us both to understand exactly why the blowdown percentage is at the number it is. If we could reduce that blowdown rate by even one-half of one percent, that would put $50k directly to our bottom line."

Vincent was visibly calmer. "Okay, yeah that would be great, but I don't know how to do that. We've been using the same conductivity limit for 20 years. I doubt it's wrong."

Tom felt the tide shifting. "I doubt it's wrong also, but given the potential cost savings, wouldn't you agree that it is worth dissecting, so we are certain. Just as importantly, maybe we'll learn something as we search for the answer."

"Sure," Vincent replied.

Tom and Vincent spent another two hours talking through the basics of boiler water chemistry and why the limits were what they were. Eventually, they stumbled upon a golden nugget. The reference document for the conductivity limits stated that the limit was established to protect the downstream condensing turbine from impingement damage.

The plant didn't have a downstream condensing turbine.

"Vincent, if we don't have a condensing turbine to protect, don't you think it would be reasonable to explore adjusting this conductivity limit?"

"Absolutely. I can't believe after all of these years this has never come up before."

After consulting with some technical experts about the downstream components that they did have and what impact conductivity would have them, Tom and Vincent were able to adjust the plant's surface blowdown rate from 15 percent down to 9 percent, resulting in annual savings of $600k.

That event was a turning point for Vincent. A few weeks later, Tom overheard Vincent asking the Operations Manager, Eric, why they were splitting the load between two boilers the way they were (60 percent of Boiler #1 and 40 percent on Boiler #2).

Eric was explaining how they had always done it that way because Boiler #1 operates at its maximum efficiency at 60 percent.

Tom's spirit welled with pride when he heard Vincent respond: "We are not relying anymore on the 'that's the way we've always done it' answer. When you say 'maximum efficiency' what do you mean? Explain that to me like I am a sophomore in high school that is starving to learn because that is all I am trying to do— learn."

A LEARNING CULTURE'S
IMPACT ON KNOWLEDGE

Our model of the Nuclear Submarine's culture of operational excellence places knowledge and learning at the foundation. All of the other elements of the culture that we are going to discuss rely on the strength of this foundation. A weak foundation can be ignored or dismissed, but eventually, the cultural elements of standards, questioning attitude, and watch team backup will falter as a result.

Standards are the elements of an organization that are supported and enforced without compromise. These are the handful of principles that each and every member of the organization must embrace. Establishing standards are simple to obtain if the only intention is to slap them on a poster and hang it in the break room. Establishing standards is much more challenging when the organization commits to support and enforce these core principles with action and enforcement at all levels of the chain of command.

In the Nuclear Submarine community, procedural compliance is a standard. When operating systems on a nuclear power plant aboard a fast attack submarine conducting missions vital to national security, operations must be performed with surgical accuracy and precision. The use of procedures minimizes human error, but no method can cover every scenario. The only protection from the situations that cannot be captured by procedures is knowledgeable operators and supervisors that understand the systems' functions, design, and operation. An organization that commits to ensuring its members are subject-matter experts will always be more prepared. Further, a culture that has been pushing its members to answer the "what if" questions during ordinary times will be massively more prepared to successfully handle unexpected events.

Questioning attitude may sound like an anti-authoritarian point of view, but as we will discuss in Chapter 4 it is a way of thinking that encourages each member of the organization to keep her brain fully engaged in the processes and situations in which she finds herself. This engagement is accomplished by asking "why?" incessantly—sometimes non-rhetorically to a supervisor or peer, but mostly importantly: "Why am I doing this?" "Why am I doing this, this way?" Adopting this type of thought process without the requisite knowledge is fruitless or more frequently inefficient.

Watch team backup (or more commonly called employee engagement) is the willingness to raise your voice to prevent a mistake from being made by someone else. The best teams are willing to support each other, but supporting each other is different than celebrating each other. Sometimes, the best support can be a forceful and public reminder that, based on a strong foundation of knowledge, what you are about to do is wrong, or worse—dangerous.

A Los Angeles class fast attack Nuclear Submarine was operating in a far corner of the world executing a mission vital to national security. The crew had been deployed for four months and was running like a well-oiled machine. Billy was nineteen years old, the son of a teacher and a farmer, and the most junior sonar operator. He reported to the boat one month before deployment. Although he had never been to New York City, Chicago, or Los Angeles, here he was thousands of miles away from his home in Lincoln, Nebraska, as a submariner aboard a fast attack Nuclear Submarine. Describing life as "difficult" for junior submariners is like saying Mike Tyson punched hard—accurate and impossible to understate. The pressures of assimilating with the submarine lifestyle are matched only by the pressure to qualify—learning the requisite knowledge to stand watch in support of the ship's mission. Billy had only recently qualified as sonar operator and was a log recorder for the ship's active sonar receiver. The submarine's operation was classified, and, although he knew he could never share the specifics of his experiences with his family or friends, he was standing his watch with vigor and professionalism that would make them proud of him.

The ship's active sonar receiver is super sensitive. Most of the signals that it detects are ocean noises and "biologics" (sound speak for fish and whales), but it is designed to identify the bearing (direction) and volume of other ship's fathometers and other navies' active sonar and weaponry. His job was to report each signal that the sensor detected to the Sonar Supervisor and then log the bearing, the frequency, and sound intensity. Yesterday, he was just studying the various frequencies of foreign navy active sonar systems and the newest signal detected was familiar to him. He reported the signal to the sonar supervisor with a raised intensity in his voice. He was sure this signal was coming from a foreign Navy's active sonar system, which can be a counter-detection threat to the invisible Nuclear Submarine. The sonar supervisor took his report of frequency, bearing, and SPL ("sound pressure level," which measures the volume of the signal). Much to Billy's surprise, the Sonar Supervisor announced to the Officer of the Deck, "Sonar, Conn, new active sonar signal bearing 123 is classified as biologics."

Billy was confused. Biologics?? That's not a fish. I was just reading about this sonar yesterday. It is the active sonar aboard the (-------- CLASSIFIED ———————).

He looked around the rest of the Sonar Room. No one blinked an eye. The Supervisor resumed scanning the displays and the other operators' eyes remain glued to their displays.

The Officer of the Deck will catch it. He'll know that's not biologics, thought Billy.

"Conn, Sonar, Aye" (an abbreviated way for the Officer of the Deck to acknowledge the Sonar Supervisor's report).

Billy spoke up, "Sup, this matches the characteristics of

(-------- CLASSIFIED ——————-). I think this is active sonar."

Without taking their eyes of their screens, the rest of the Sonar Operators laughed in unison. One operator spouted, "NUB" and shook his head. (NUB is a playfully derogatory term reserved for junior personnel that stands for "non-useful body.")

The Sonar Supervisor replied, "There is no way a (-------- CLASSIFIED ——————-) is operating in these waters. It has to be biologics."

Billy paused. He knew he had much to learn; he knew he was the junior sonarman, but he also knew that frequency exactly match (-------- CLASSIFIED ——————-) and thought that the Supervisor should at least let the Officer of the Deck know that it was possible that (-------- CLASSIFIED ——————-) was operating nearby.

Billy held his ground, "Aren't we supposed to report to the Officer of the Deck that it is possible that a (-------- CLASSIFIED ——————-) is operating in the vicinity."

The 15-year salty veteran and the 4-month veteran held each others' gaze for a moment. The Supervisor admired the kid's guts and knew, deep down, that he was right.

"Conn, Sonar, be advised that the signal bearing 123 previously classified as biologics is a possible (-------- CLASSIFIED ——————-)."

Guess what the signal bearing 123 turned out to be? If you guessed (-------- CLASSIFIED ——————-), you are correct.

Billy's dynamic watch team backup, which we will discuss more in the Chapter 5, was possible because of his willingness to speak up, but his knowledge gave him the confidence and information to do so.

RECOMMENDATIONS FOR BUILDING A LEARNING ORGANIZATION CULTURE

We can imagine how difficult it would be to ask all of the professionals around us to prove themselves—again. Asking your Regional Engineer with 30 years of experience to re-take the Professional Engineer's examination would undoubtedly result in an uproar likely followed by his resignation. We understand this and are not suggesting it. However, we are suggesting that evaluating the "intellectual integrity" of your organization's professional knowledge is a necessity because it will impact the organization's ability to thrive when others falter.

There is no particular mechanism that we are aware of to perform this evaluation; however, we present the following for your consideration:

Ask "What if?" often. We discussed this at length when we discussed Knowledge as a way to explore the depths of your team's imagination and knowledge. We're now adding a slightly different intent to this mechanism. It is often hard to identify intellectual integrity after only the first response. Your team may have said or heard a version of that reply countless times in their careers. But have they ever dug deeper? Our "what ifs" are not about identifying how deep below the surface a person's knowledge goes; our goal is to discover how does a person respond when he does reach the bottom. Many professionals refuse to say "I don't know." Is this the person that you want representing your organization when a situation goes off the beaten path? Is this person an exception or the rule in your organization? Unless you are running a startup, you may not know what type of culture conditioned the organization in years past. Identifying the "I will not say that I don't know" crowd is not a witch hunt or judgment; it is an evaluation. Your goal should be to foster an environment that allows the members of this group to drop their defenses, take a breath, and know that they are in an organization that expects and accepts, "I don't know, but I will find out" as an answer.

Encourage cross-training. We live in a specialized world. We must. Technology has grown overly complex, advances at a staggering rate, and the competition is fierce. It is highly unlikely that I'll be able to maintain competitiveness if my computer coder is also my marketer, my financial analyst, and my attorney. Outside of a one-man, bootstrapped startup company on day one, that example is preposterous. However, there is real value and a robust return on investment in cross-training. Unfortunately, it is nearly impossible to measure the return on investment that cross training provides; however, when the accountant can discuss the operational lessons and the forklift operator understands the company's basic finances, real value is added to the organization. If you've experienced it, you know it. Further, the process of allowing for cross-training through the organization provides a tremendous opportunity to seek and identify hidden talent. Your financial analyst might be a budding world-class marketer, but if you never expose him to the core elements of the company's marketing plan, that may never come to light.

Seek out simplicity. When you don't understand a concept or answer, request that your team explain the concept such that a fifth-grader could understand it. (We stole this idea from Albert Einstein and Richard Feynman—both physicists used this litmus test to explore if someone understood something.)

This litmus test is an interesting exercise in evaluating both intellectual integrity and communication. Every industry develops a vernacular that insulates its members from the "outside" world. However, members from different industries communicate with one another every day— few projects can be accomplished without support from at least one other industry. So, insider vernacular can impede effective communication. Consider a building owner who is requesting a design from an engineering firm to install three flux capacitors in the basement. The business owner asks a simple question and receives a load of complex techno-babble in response. The owner asks again and gets a similar result. The communication barrier that began as industry-to-industry translation quickly bleeds into the realm of (dis)trust. We discuss this frequently as a challenge in all of our industries, but it is usually considered from strictly a communication perspective. But perhaps it is simply inadequate knowledge hiding in the bushes of complex terms.

Read about your industry—voraciously. News, trends, events. Share information that you find interesting with your team. Share information that you don't understand with your subject-matter expert and ask them to explain it to you. Your passion will be infectious, and you will, with genuine and persistent effort, reinvigorate people's desire to learn more and increase the organization's intellectual integrity.

Learn – together. Take a MOCC (Massive Open Online Course) with one or more of the group. (A catalog of MOCC course can be found at https://www.mooc-list.com.) Carve out an hour or two a week to listen to the lectures and go through the coursework together. This is another great way to evaluate and strengthen your group's intellectual integrity. It is our observation that there is a sweet spot between Google's corporate "let's give the children playtime" culture and a rigid environment that forgets that intellectual growth does not need to dead end. But again, this isn't just about knowledge; it is the integrity of knowledge. If you lead one of these courses, don't feel restricted by the curriculum—if (when) something interesting is discovered, follow it. With your actions, show the group your willingness to say, "I don't know this. Can we back up and talk about it?"

Foundations of organizations, like buildings, serve to strengthen the structure especially when the unexpected stresses occur. The components of the Nuclear Submarine's culture of excellence are only as strong as its knowledge and its commitment to knowledge as evidenced by the learning culture that exists. As we explore the other components of the Nuclear Submarine's culture of operational excellence, keep this fact in mind. As we bury ourselves neck deep in financial statements, marketing plans, and operational performance, the potential to lose sight of the basics exists. The consequences of doing so can be damaging and long-lived while the opportunity to increase and maintain the organization's sustained superior performance is within reach by refusing to lose sight of those basics.

CHAPTER 3

STANDARDS

In our metaphorical building model, standards are the doors. Everyone that enters the organization does so through the door of standards. These are non-negotiable attributes that are applicable to the night cleaning crew and the CEO. Standards can also be the back door, where people are shown the exit if they are unable or unwilling to meet the organization's standards.

Standards are organizational attributes that the organization holds in the highest regards. Optimally, an organization's leadership would develop standards that are the most fundamental to the organization's success. The standards in your organization will be different from other organizations and may change from time to time. The specifics of the standards are vitally important but are secondary to the manner in which an organization communicates, enforces, and executes. Without laser-like focus on each of these processes, an organization is likely to be left with the shadow of standards that look pretty on the poster in the break room but do little to impact the organization's path to excellence.

We are committed to safety, professionalism, and integrity sounds good, but the devil is in the details. Consider a standard of safety expressed as "Our standard is zero injuries or accidents." That is not a standard—it is a goal. It does not effectively get to the heart of the matter.

NUCLEAR NAVY: PROCEDURAL COMPLIANCE

The most unique standard of Nuclear Submarines is our standard of procedural compliance. Successful standards are communicated in a manner that allows the standard to be met. In many organizations, standards are believed to be one thing in the executive conference room, but are applied in a much different manner in the welder's shop. The manner in which the standards of procedural compliance is enforced in the Nuclear Submarine community is NOT: "We use procedures for everything that we do." Our Nuclear Submarine brethren reading this probably just raised their eyebrows in disbelief. This axiom, although more comfortable to say, is not, in fact, the standard that we demand and perform at all levels in the chain of command. It works well as a training aid to the junior personnel, but it is the equivalent to saying: "Everyone will give 110 percent effort on this project." It sounds inspiring but also unrealistic and impossible in practice.

A more accurate expression of our procedural compliance standard is: "We follow procedures for all non-emergency operations, and we follow immediate and supplemental actions in procedures for equipment casualties, except when operating conditions require a departure from procedures, in which case operator knowledge and experience is relied upon to take the correct action to restore the submarine to a safe operating condition. We also acknowledge that there are always procedures or a source document that provides guidance or direction for all normal submarine operations."

This standard may seem more convoluted than the first, but it is more accurate and, therefore, more enforceable and executable. This standard does not create the hypocritical delta between what is acceptable in the Commanding Officer's stateroom at 0900 and also in Engine Room Lower Level at 0300.

It would have been much easier to write this chapter if we just stuck with the simpler "We use procedures for everything all of the time." Unfortunately, this is just not true, and it is also one of the most damning mistakes that organizations make when establishing standards—specifically, setting unattainable standards. This creates an uncomfortable hypocrisy, which, although it is felt by everyone in the organization, it is addressed and discussed only behind closed doors in a whisper.

Standards must be obtainable. Unrealistic standards lead to no standards.

However, the Nuclear Submarine force does have a reverence for procedures that would stun even the most meticulous civilian. If you ever have the opportunity to tour a Nuclear Submarine, be sure to use the restroom. You will be surprised to see that there is a posted procedure on how to use the restroom. Granted, the toilet isn't a Home Depot Memorial Day sale model, but still . . . a procedure to operate the toilet? Yep!

So, what's the deal with these procedures? (Asked in Jerry Seinfeld's nasal tone.)

Allegiance to procedures is not, as some assume, a negative reflection on the operators who use them. In fact, as you will read later in this chapter, the exact opposite is true. Consider your morning routine. How would you perform that routine if there were dire consequences for forgetting something? Assume, for maximum dramatic effect, that everyone you cared for would lose an arm if you left out or screwed up a step in that routine. Have you ever left the house without pocketing your wallet, matching your socks, or taking your daily multivitamin? What would you do differently if it was imperative that your routine was exactly correct one million times out of one million times?

I suspect that you would either refuse to leave your home or implement a procedure and a checklist.

When something has to be done exactly correctly every single time, procedural compliance protects us from our worst day or a momentary lapse of reason. On a Nuclear Submarine, following the procedure is just the beginning. The manner in which the procedure is executed is the part that many people have a tough time wrapping their brains around. Before you try to understand the following example, bear in mind that operations on a nuclear power plant on a submarine hundreds of feet beneath the ocean's surface are perfect examples of operations that we must strive to do correctly one million out of one million times.

All operations of nuclear systems are directed from a small control room within the Engine Room called the Maneuvering Area. The Engineer Officer of the Watch (EOOW) is responsible for all Engine Room operations and directs these operations from Maneuvering. The Engineering Watch Supervisor (EWS) directly supervises these operations from the Engine Room. These two supervisors communicate with each other through the use of phone talkers—individuals whose only responsibility for the evolution is to send and receive messages through the use of the Engine Room's sound powered phones (telephones). Then there are the operators who actually do the work.

Assume we are executing a procedure that will be adding water to the primary coolant system to make up for the losses associated with our daily sampling. Further assume, the first step of many steps in the procedure is to open a valve identified as "charging 42" (a valve in the primary charging system that is uniquely identified by its number and written as CH-42). The following illustrates the standard of how that valve gets opened.

- EOOW (to the Maneuvering Area Phone Talker): "To the EWS, open charging 42" (The number 42 is spoken as "four two.")

- Maneuvering Area Phone Talker: "To the EWS, open charging 42." (This is called a "repeat back" which is used to reduce the possibility of a miscommunication.)

- Maneuvering Area Phone Talker (to the EWS' Phone Talker): "To the EWS, open charging 42."

- EWS Phone Talker: "To the EWS, open charging 42, aye." ("Aye" is Navy speak for "I understand.")

- EWS Phone Talker: "EWS, from Maneuvering, open charging 42."

- EWS: "Open charging 42, aye."

- The EWS will then reference the procedure to confirm that he concurs this is the correct order.

- EWS: "Valve operator, open charging 42."

- Valve Operator: "Open charging 42, aye."

- The valve operator will then point to CH-42 and mimes (without touching the valve) the direction he intends to rotate the valve. After doing this, he will pause for a second or two before opening the valve. This method (called "point and shoot") allows the EWS to be certain the valve operator is going to operate the correct valve in the correct manner with enough time to intervene if this is not true.)

- The valve operator then opens CH-42.

- Valve Operator: "EWS, charging 42 is open."

- EWS: "Charging 42 is open, aye. Phone Talker, to Maneuvering: charging 42 is open."

- EWS' Phone Talker: "To maneuvering, charging 42 is open, aye. Maneuvering, from the EWS, charging 42 is open."

- Maneuvering Area Phone Talker: "From the EWS, charging 42 is open, aye."

- Maneuvering Area Phone Talker: "EOOW, from the EWS charging 42 is open."

- EOOW: "Charging 42 is open, aye."

Now we move on to second step in the procedure, which is executed in the same way. And . . Yes, this example accurately depicts the process of executing a procedure in the Nuclear Submarine community.

All of that to open a valve?! *Seems highly inefficient,* you may be thinking. There is no doubt this is inefficient, but when it comes to the operation of nuclear systems, effectiveness is king, not efficiency.

This example serves to highlight the meaning of "procedural compliance" on a Nuclear Submarine—it's about the manner in which the process is done as much as the action itself. Re-read that dialogue and try to envision a circumstance under which the incorrect valve would be opened. The use of Phone Talkers, repeat backs, the "point and shoot" process, and the

EWS' independent check of the procedure are all in place to make it virtually impossible for the incorrect valve to be operated.

IMMEDIATE ACTIONS

The Nuclear Submarine community does not require operators to memorize procedures. In fact, memorizing procedures is frowned upon. No one's memory is perfect, and, as we discussed earlier, we are striving for a process that asymptotically approaches perfection. However, sometimes there is no time to open a book. The perfect example is one's response to a fire. Responding to a fire on a Nuclear Submarine by opening a book is unthinkable. There is nothing scarier to a submariner than fire. Even a modestly sized fire will fill the entire space of submarine with thick black smoke in a matters of seconds. Our response to fires is instantaneous and every second counts—truly. Therefore, every submariner has the immediate responsive actions to a fire memorized. This is true for every casualty that requires immediate action to save lives, the ship, or the reactor plant such as fire, flooding, steam line ruptures, or radioactive spills.

There are no B+ grades when it comes to memorizing immediate actions; there is one standard and one standard only—complete memorization. Both of us can attest to the fact that when a real casualty occurs the mind shifts into an entirely different mode of operation. Time slows down Matrix style; the ability to think and reason is impeded; the urge to "fight or flight" is real. The only acceptable response—to fight—stems from the muscle memory of our memorized immediate actions.

Immediate actions lend themselves well to dangerous situations as described, but it would be reasonable to extend this process beyond physically threatening scenarios into the realm of operational threats. Examples of when opening a book of approved standards would not be appropriate or efficient in anyone's work world would include when dealing with an upset customer, a belligerent employee, or a delay in the critical path of a project.

CULTURE OF PROCEDURAL COMPLIANCE

The standard of procedural compliance in the Nuclear Submarine community is comprised of more than the procedures and the methodical process by which they are carried out. There are two additional components of this standard that are worthy of discussion. They are preparation and thinking operators.

Preparation

With few exceptions, doing something fast on a submarine has no virtue onto itself. Think back to the hypothetical morning routine we discussed. Expand that such that it applies to every member of your family. So, if anyone in the family misses a step in his or her morning routine the rest of the family loses a limb. In that bizarre world, can you envision congratulating your children on how quickly they left the house in the morning? Probably not. So too, on a Nuclear Submarine, although sharply efficient operations are the norm, efficiency never trumps effectiveness. To that extent, procedures on a submarine begin with a pre-evolution brief. These briefs vary in length, content, format, and people in attendance, depending upon the complexity of the evolution.

The brief is typically led by the supervisor who will be directly overseeing the operation or maintenance and is attended by the people who will be participating or supervising all or parts of the operation. The pre-evolution brief is designed to give everyone an overview of what we are doing, why we are doing it, and how we will do it. More specifically, the procedure that will be used is reviewed, the method of communication between the different players is discussed, and questions are asked to address anyone's concerns and also to test retention.

There are three additional elements that every strong pre-evolution brief will include: precautions, initial conditions, and lessons learned.

Precautions. Precautions are the list of items that precede the procedure, and this list grows with time as new lessons are learned. No procedure can possibly capture every element of every operation. Precautions provide the team a list of things to consider and remember prior to commencing the operation. An oversimplified example would be a reminder that steam pipes are extremely hot and should not be touched without protective gear. A more realistic (although technically fictitious) would be: "When moving the Diesel Generator Bull Bearing switch from "off" to "start," be sure to rotate the dial clockwise to avoid moving the switch through the "self-destruct" position."

It is often said that the precautions are written in blood. Although, slightly melodramatic, it is true that precautions are typically included in procedure revisions after an unexpected consequence, a mistake, or a near miss occurs. These precautions are included to prevent their recurrence. Submariners are reminded many times in their training that even though some precautions are inconspicuous or even seemingly obvious ("don't grab a hot steam pipe with your hands"), someone just as smart as you made that exact mistake. Stay humble and review each precaution carefully, and, remember, there is a very interesting story behind each one.

Initial conditions. Initial conditions are a list of system line-ups and parameters that the procedure assumes exist. If these initial conditions are not met, the procedure should not be performed unless they can be met or the organization's leadership decides how to alter the procedure to account for this variance.

Let's play a quick game to emphasize the significance of the initial conditions. Take a moment to think through how you would write a procedure to:

1. Replace a flat tire
2. Open a bank account for a new customer
3. Make a peanut butter and jelly sandwich

Would your procedure work if:

1. The car was parked on a 30 degree incline?
2. The new customer was blind?
3. The bread was frozen?

Maybe or maybe not, but it is clear you cannot write a procedure to cover every possible case. In many cases, a small deviation in initial conditions could cause damaging consequences if the procedure was attempted. Since we can't cover every set of possibilities, we can ensure that we precisely describe the conditions for which the the procedure IS written. Then, it is in the hands of the operating team to validate if their present condition does or does not meet these same conditions.

Lessons learned. Capturing lessons learned after an operation or evolution is vital to the long-term success of an organization. These lessons are also painful to gather. In most organizations, once an operation is over—it is over. The idea of spending more time on a finished evolution seems like a waste of time, especially since there always seems to be a new, shiny object taunting us for attention.

This response is natural but also very selfish. The future success of the organization depends largely upon its abilities to learn from the organization's previous mistakes. Often, we will go through a process that was riddled with mistakes from start to finish. The process was so painful that a person is likely to think, "I don't need to write any of that down; I'll never forget the lessons from these battle scars." Although that may be true, the organization doesn't have the benefit of those battle scars unless the mistakes and lessons learned are written down.

To highlight the importance of capturing lessons learned, they are included in the pre-evolution brief. When the supervisor is preparing his pre-evolution brief, and there are no recorded lessons learned for this evolution, he will be forced to seek out information about the last time this evolution was performed. The option of saying, "And there are no lessons learned to share" is not a viable one unless the procedure is so infrequently performed that there really isn't any organizational knowledge— which, in itself, should bring to the team to a more heightened state of alertness.

> *Learn all you can from the mistakes of others.*
> *You won't have time to make them all yourself.*
> ~Alfred Sheinwold (Master American Bridge Player)

There are other advantages to the pre-evolution brief that are less direct. For example, it gives the supervisor an opportunity to observe the entire team and look for intangibles that go beyond procedural compliance. Why does Tom look so exhausted? Did he sleep today? Why is Senior Chief Klein so much crankier than usual? Will this negatively impact his ability to supervise?

Some pre-evolution briefs reveal that the team is not ready to conduct the operation. This is an element of the pre-evolution brief that is very contextual. There are some operations where we consider the pre-evolution brief to be more of a test than a brief. For these operations, there is an expectation that the players already understand their roles and the procedure. A good example of this is a reactor startup brief. The Nuclear Submarine community conducts hundreds of reactor startups a year without incident. However, starting up a nuclear reactor places the reactor in its most unstable condition—the transition from where there are virtually no nuclear fission reactions to a condition where there are exactly enough fission reactions to keep the heat output constant and sustainable (critical) but no more. Arguably, just as sensitive is the process of bringing the high pressure and temperature steam created by these nuclear reactions into the engine room. A reactor startup brief is not a learning experience; it is a certification. If the Commanding Officer gets even a sniff that the team is not prepared, the start-up will be delayed until the team can prove that it is ready.

Thinking Operators

Here's a horrible idea. Tell a submariner that he has an easy job because "all you guys do is follow procedures." Go ahead, but don't say you weren't warned. The standard of procedural compliance requires thoughtful operators whose minds are constantly thinking one step ahead and whose senses are constantly observing their surroundings for indications that the systems are responding as they should be . . . or not.

The best example of the requirement of thinking operators is a process called "verifying system response." Our previous example of the communication required when opening a primary valve did not include an element in the process that is just as fundamental to our standards of procedural compliance as the communication process. Prior to opening or shutting a valve, starting/stopping a pump, moving reactor control rods, or adjusting plant voltage or frequency, the operators are expected to identify the expected system response. In our example of opening the valve, CH-42, the operator and all others involved in the procedure are expected to know what is going to happen when that valve is opened. Should the operator hear a flow noise? Should water flow through a downstream valve? Should pressure increase at a particular point in the system? Should reactor power or steam flow be impacted? Recall, the procedure being performed was adding water to the primary system. This is multi-step procedure with many valve and pump manipulations, so being able to identify system response for each step requires the operator to be a subject matter expert on the reactor plant and the charging system's impact on the overall reactor plant operation.

The expectation doesn't stop there. What if the operator doesn't see/hear/feel the response he expected? What should he do? Shut the valve? Maybe or maybe not. It depends on variables that exceed the technical scope of this chapter and book. Suffice it to say, being able to open a valve because a procedure tells us to is a galaxy away from our standards of procedural compliance.

Although sore subjects for proponents of nuclear energy, the mishaps at Three Mile Island and Chernobyl were a direct result of failing to respond to system responses that were unexpected. Assuming the best or worse, turning a blind eye to a system response that is not expected is damaging in any organization and potentially life threatening in industrial environments.

Another element of the procedural compliance standard is the expectation for operators and supervisors to evaluate the procedure itself while using it. None of our procedures were divinely inspired. Each one was written and reviewed by fallible human beings. As a natural consequence of this irrefutable fact, procedures will have errors in them. The procedural compliance standard is sustainable because there is a feedback loop that works to continuously improve our procedures. If an operator or supervisor identifies an error, he is expected to use his system expertise to decide what to do in real time.

Typically, the correct response is to restore the system to a stable condition, stop the evolution, and then inform the chain of command. Procedure errors take on different shapes and sizes. Some are small and require only an administrative correction: "Open valve CH-42" is a typographical error. Others are substantial and indicate that either the procedure is significantly flawed or there is a problem in the plant's operations. Looping back to our "open CH-42" example, again: What if the operator attempted to open CH-42, but it was already open. This is significant. If it was expected to be open, the procedure would have directed the operator to "check open CH-42." It is possible that the procedure is wrong, but it is also possible that there was a reactor system valve that was "out of position," which is an untoward event that would require a very thorough investigation (known as a "critique" as described in Chapter 7).

THE BENEFITS OF
PROCEDURAL COMPLIANCE

In the Nuclear Submarine community, there is a shared and unequivocally staunch respect for written procedures. But why? What are the benefits of this standard? What are the benefits to non-military organizations? Here are the most important benefits that any organization can expect to gain by establishing the standard of procedural compliance.

Consistent Output

The massive success enjoyed by Starbucks and McDonalds is not because they offer the best coffee and food. Instead, they offer a known quantity. The entire experience (menu, ordering process, product, taste) is identical in Seattle and Tampa Bay. This consistency is not accidental. It is a result of establishing and then enforcing a standard of procedural compliance. Consumers reward consistency. Procedures create consistency.

Workforce Flexibility

Procedures increase the effective size of an organization's labor pool. If the same procedures are implemented and enforced throughout the organization, the organization can more easily shift personnel from one location to another without transition pains.

The Power of the Pen

Directions and guidance become immediately more powerful when written down. Why? Who knows, but it's true. We view written words with a reverence that is much more powerful that the spoken word. Not convinced? Consider the following.

You are a junior project manager for the installation of commercial security systems. Compare these two methodologies for assigning a new project.

1. Your boss, Mike, stops by your cubicle on Monday morning and says, "Hey, Shirley. We locked up the contract on the facility on Olive and 3rd. Make contact

with their facilities director and get started on the project schedule. Let Sara and Kyle know they'll be working with you as well.

2. You receive an internal letter or memo that reads as follows.

Shirley,

You are an excellent project manager. The last two project installations that you managed were performed expertly. Recently, we discussed the opportunity to increase the depth and breadth of your weekly status reports on your next project. We successfully signed an installation contract with Megabank Building on the corner of Olive and 3rd Street.

This project is yours. YOU are in charge. Sara and Kyle will form your team and will report directly to you. I have full faith and confidence in your leadership and professionalism.

YOU are the leader and owner of this project, and I am here to support you in anyway that I can during the project.

Sincerely,
Bill Lumbergh

Which is more powerful?

Reduction of Workplace Adversity

When an employee makes a mistake, bosses get upset. Some things in life just "are"—this is one of them. However, when there are no procedures or policies in place, the post-mistake discussion is often very confrontational and embedded with "he said, she said" discussions. Invariably, supervisors lament that their employees "just don't listen" and workers complain that their supervisor "blames me for everything." When there is a procedure in place, supervisors can use the procedure as the "bad guy."

Let's take a look at this situation: A contract with ZBest Contractors was signed without a performance guarantee.

Here is the conversation without a procedure in place:

Mike (Supervisor): *I can't believe you didn't require a performance guarantee. How many times have we talked about this?*

Amanda: *Mike, we've signed contracts without performance guarantees in the past. In fact, the last contract that you and I worked on didn't have a performance guarantee.*

Mike: *Ugh! Amanda, that was a supply contract, this is an installation contract. We never agree to an installation contract without a performance guarantee.*

Amanda: *I don't know, Mike. It seems like the rules are always changing for me.*

Mike: *Well, it seems to me that you have a problem following instructions. Further, you have a tendency to blame everyone but yourself when you do make a mistake.*

Et cetera

Here is the conversation with a procedure in place:

Mike *(Supervisor): Amanda, our newest installation contract was signed without a performance guarantee. Why?*

Amanda: *I didn't think one was required.*

Mike: *Hmm. Let's take a look at the procedure for installation contracts. Step 12 states to ensure all installation contracts contain performance guarantees, and it lists the minimum requirements of that guarantee. I'm confused. Did you not reference the procedure?*

Amanda: *I didn't.*

Mike: *Why not?*

Amanda: *I was in a rush, and thought I had had the rules memorized.*

Mike: *No one is perfect, which is exactly why we have the procedure. It is a tool we created for your own protection. Our standard is to reference the procedure for all administrative client work. Do you understand this standard?*

Amanda: *I do. It won't happen again.*

Improvement of an Organization's "Communication Lifecycle"

Communication lifecycle is the effectiveness of maintaining the veracity of communication over time. For example, consider an organization that has new inventory-management software installed. At the time of installation, all of the users attended a three-day training program provided by the software vendor. These users were trained well and operated the system to its maximum capacity nearly flawlessly. Fast forward five years. How proficient are the operators now? The ones that were part of the original training group and those that were hired since? The answers to these questions are a reflection of an organization's ability to obtain, retain, and transfer information over time. Written policies and procedures allow organizations to effectively do so. If the information is not captured this way, the information will be retained and transferred via word of mouth. We all learned in kindergarten just how effective this method is.

"Standing orders" are another example of using written guidance to improve the effectiveness of an organization's communication. Standing orders are written directions from a supervisor to the group she supervises that cover a wide range of situations. For example, on a Nuclear Submarine, the Commanding Officer maintains "Commanding Officer's Standing Orders," which contains guidance to the entire boat and more specifically the supervisors. One of the more effective and unique elements of these standing orders is providing very clear guidance about what actions require the Commanding Officer's permission. Further, it contains clear guidance on when to inform the Commanding Officer on a particular item. This eliminates the uncertainty about how information should flow to the Commanding Officer.

Written policies and procedures have so many advantages, but, yet, these advantages remain unclaimed in so many organizations. In fact, we have observed many environments above the ocean's surface that hold abject aversion to written policies and procedures. Nuclear Submarine veterans are usually flabbergasted by this aversion, but the reality is that most organizations do not embrace procedural compliance as a cultural standard. Many industrial organizations do something even worse—they claim to expect procedural compliance but frontline supervisors do not enforce this standard. As a result, a hypocrisy develops that is cancerous to the organization's integrity. There are few working environments that are more frustrating than one with stated standards that are disregarded. Not only does the organization lose the opportunity to create the cohesion that comes with accepted and enforced standards, but it risks developing a culture of suspicion and disrespect of organizational standards and expectations.

Why does this hypocritical situation occur so frequently when it comes to the use of procedures?

There are many reasons why we observe resistance to procedures—many of which would be better addressed by psychologists than by us, but here are some obstacles that you may have to overcome to establish a standard of procedural compliance.

Fear of being cornered. Often, midgrade/senior personnel are averse to procedures because they fear that they are being setup for a "gotcha" moment. They fear that the procedures will be used as a tool against them when their work does not meet the organization's standards. "I've been working without procedures for 20 years, and now I am going to be treated like the guy on fries at McDonalds?" The objective is to convince this group that the exact opposite is true. Procedures can be used as a tool for workers to be confident that their work meets company standards.

They don't know how. Some people do not know how to use procedures. There is no kind way to say this. This is observed most often with junior personnel and positions. This is a much easier resistance to overcome because people can be taught how to use procedures and participate in the procedural compliance standard. But, because nothing is easy, the people who need this training will also need convincing that they do. Further, many supervisors overlook this weakness as a possibility when attempting to implement the use of written procedures. They incorrectly assume that they are dealing with a "won't" situation, when they are actually dealing with a "can't" situation.

Procedures shine a light in dark corners. There is a gap between the way management expects an organization to run and the way it is actually run. There is probably no exception to this, even in organizations with procedural compliance standards, but the difference is the size of the gap. Procedures expose this gap plainly, and this makes everyone that previously abetted the gap nervous.

No one likes to write procedures. Following procedures and earning buy-in from the organization is challenging in and of itself, but getting them written is a whole other ball game. No one likes to write procedures, and few are good at it. Therefore, it is common to receive resistance about the need for a procedure from a person that knows she will have to write it. As we discovered while writing this book, Ernest Hemingway was correct when he quipped, "The first draft of anything is s**t."

Overcoming these resistance points starts with the awareness of their existence. Next is an open dialogue that acknowledges these concerns and emphasizes the benefits. It is important to emphasize the amnesty of the process—if you discover something that was not being done correctly in the past as the organization establishes procedures, there are no negative professional consequences. The transition is not a witch hunt; it is an excellence hunt.

PROCEDURAL COMPLIANCE MINDSET

We've discussed the procedural compliance standard in terms of the contents of the procedures (initial conditions, cautions, steps), and the manner in which they are executed (pre-evolution briefs, deliberate communication, and action). We discussed the benefits of establishing these processes as a standard. The standard also breeds a second-order consequence that is subtle but powerful: the establishment of a mindset that can be applied even when there is no policy or procedure.

In the first section of this chapter we made the point that even in the Nuclear Submarine community, we don't use a procedure for everything all of the time. There are times when operations need to occur at a pace that does not allow for a step-by-step following of a procedure. There are other circumstances, although rare, when supervisors make thoughtful and deliberate decisions to proceed without a procedure or to deviate from one. In other instances, the evolutions are routine and low risk, and we trust operators to perform these duties correctly without the use of a written procedure in hand—pumping the bilges of the Engine Room, for example.

However, the procedural compliance standard encourages a mindset that creates a consistent approach to problems, evolutions, and operations. The procedural compliance standard is the equivalent Mr. Miyagi's "Wax on, wax off." For example, an operator in a manufacturing plant notices that there is excessive vibration in a grinding wheel. In an organization with procedural compliance standards, the operator is likely to take a moment to consider the "initial conditions" (what is the existing plant lineup?) and the "cautions" (What are the risks associated with the current condition?). After preparing himself with these thoughts, he would communicate the facts to his supervisor using deliberate and effective communication. The supervisor would assess the need for immediate action versus a more deliberate plan. An observer would be able to match their response to the various elements of procedural compliance.

THE COST OF HYPOCRISY

At a power plant teetering on profitability, Tim (the Plant Manager, PM), Kevin (the Assistant Plant Manager, APM), and Marie (the office administrator, OA) assembled at the office doorway of Adam, the General Manager (GM). The news was bad and, although the GM had proven to be a reasonable person in his first year at the plant, they had not seen how he would react to news that was this bad before. There was strength in numbers as they gathered awkwardly in doorway waiting for Adam to invite them in.

After getting off the phone, Adam waved them in. "Uh oh. The whole gang is visiting. This can't be good. What's up?"

The PM wasted to no time. "Sir, we just discovered an improperly calibrated steam meter gauge at the city's conference center. The gauge has been reading incorrectly for at least three years."

The GM listened intently with a poker face. He knew that lead-in was only beginning. "Okay, now cut to the chase."

The OA piped up, "We've been underbilling the city conference center for three years."

The GM's stomach was immediately thrown into knots, but he successfully maintained a calm demeanor. "Marie, what's the total damage? How much did we underbill?"

"Approximately $500,000."

The PM, APM, and OA's heads were literally lowered in shame, not unlike school children sent to the principal's office. The plant had been struggling for nearly a decade and ended the previous year in the red; a $500k underbilling was not just a mistake, it was an abysmal failure. In response to this news, no reaction short of physical violence would have been considered an over-reaction. Every cell in his body longed to lash out explosively. However, the GM saw before him three of the most well-intended, hardworking and professional people he had ever known. *Don't shoot the messenger,* he reminded himself. *Stay calm.* He then calmly asked, "What do we know at this point?"

Kevin blurted, "It was Josh. He installed the meter incorrectly three years ago and was the one who calibrated the meter for the last three years. You have to fire him. This is the last straw."

Josh was an instrumentation and control (I&C) technician whose performance and professionalism had been satisfactory, but no more, for the duration of his seven-year tenure with the company. The past year was a different story. He struggled to stay in the good graces of the chain of command. He lost his temper at a worksite in the presence of his boss, subcontractors, and representatives of one of the company's customers. These unprofessional outbursts continued throughout the year. The chain of command had urged Adam to terminate him on several occasions. However, his poor performance and attitude coincided with a reorganization that Adam implemented, which changed his boss. Wes, his new boss, was a peer of Josh's for seven years. Adam suspected the problem was a two-way street. Wes had never held a supervisory position and was still learning basic leadership and management skills, especially conflict resolution and communication. As a result, Adam had chosen to allow Wes and Josh more time to work through their growing pains rather than take the advice of the chain of command and terminate Josh's employment.

"Tim, do you agree that this was entirely Josh's fault?"

"Absolutely. He installed the meter incorrectly and then signed off on a calibration for three straight years. This is gross negligence."

"Do you agree that we should terminate Josh's employment immediately?"

"Without a doubt."

Adam exhaled loudly. "Okay. I'll talk to Human Resources and make them aware of my intention to release him. Tim, please suspend Josh without pay. Tell him that we will contact him in a few days."

"Will do. Anything else, sir?" Tim responded.

Adam straightened himself up in his chair. "Umm, yeah. Josh's employment is not our biggest problem here. Of our 102 customers, how many of those gauges had their annual calibrations performed by Josh."

Tim turned to Kevin, who was previously the I&C Supervisor before assuming the duties of Assistant Plant Manager one year ago.

Kevin responded sheepishly. "All of them. Josh has been our calibration guy for years."

Adam could feel his blood boiling and head pulsating as a harbinger of a painful tension headache. *Don't shoot the messenger. Be the eye in the center of the storm,* he advised himself.

Adam responded calmly but with a volume just above a whisper that the group interpreted, correctly, as a sign of his restrained frustration. "Tim, at this point, we must assume that every customer gauge is reading incorrectly. Every single one."

Marie, who handled the billing, piped in with the best of intentions, "Adam, I haven't seen anything odd in our billing numbers. The other sites seem to be reading correctly."

"Marie, did the convention center bill look odd to you at any point in the last three years?"

"No, sir." Marie understood Adam's point and now regretted saying anything.

"Tim. How long does it take to calibrate a steam flow meter?"

Kevin answered for Tim, "Two hours."

"Okay, well we have 204 man-hours of work ahead of us. Tim, please work with Wes and put a plan together about re-calibrating each of our meters. Of course, let's start by prioritizing the larger customers. In fact, I would like a report when our top ten customers have had their meters calibrated. Additionally, I would like an immediate report for each meter we find that was calibrated incorrectly."

"Yes, sir," replied Tim.

"Gentlemen, thank you for your candor, and I appreciate, in advance, the amount of work that this is going to take to unscrew. Thanks."

The group was out of Adam's office and halfway down the hall, when Adam realized he had another question. "Kevin!" he yelled.

Kevin returned to Adam's office, "Yes, sir?"

"Do we have a steam flow meter calibration procedure?"

"Yes we do."

Adam had difficulty sleeping that night. Underbilling a customer half a million dollars was going to be quite a large black eye for the plant. But that wasn't it. The plan to terminate Josh weighed heavily on his mind and heart. Something didn't feel right. Josh wasn't a superstar tech, but would he really install and then calibrate a gauge incorrectly each year for three years? There had never been a documented case of Josh's poor technical performance. His problems related to his attitude not his work. Maybe the procedure was flawed? Maybe he wasn't using the procedure and was trained improperly years ago? Adam feared that he was making an emotional decision. He recognized that when a mistake of this magnitude is made there is an instinctual need to blame someone, but was he sure that Josh was deserving of this blame? What if it was someone else who calibrated the gauges? Would his response be the same? His spidey sense was alight with the suspicion that there was more to this story.

The next day, Adam asked Kevin into his office.

"Kevin. I am going to ask you some questions that were haunting me last night. First, let me just say that I know that you feel a sense of responsibility and probably professional embarrassment about this situation since you were the I&C Supervisor when these mistakes happened. I get that. However, I need you to know that I wouldn't have promoted you to your current position unless I was fully confident in your talents, knowledge, and potential. Nothing about this situation is going to change that. That's important for you to understand because getting to the truth on this situation is going to require your brutal honesty. Does that register?"

"Yes, sir," Kevin answered with a blend of confidence and apprehension.

"Does the I&C group have a standard of procedural compliance?"

"Umm. Not for everything and probably not to the extent that they should. I know Wes is working on fixing that."

"Do you think Josh understood that he was expected to follow the calibration procedure or do you think he fell into a trap of just doing the process that he thought was correct?"

"He should have known."

"Why?"

Kevin searched for answer and then settled for the best he could, "I have seen Josh leave the plant to do calibrations and the laminated procedure was always in his calibration bucket."

"Hmm."

"Okay. Let's do this. Grab the calibration procedure and equipment. You and I are going to calibrate a steam meter."

"Yes, sir. No problem." His confident words belied the expression on his face.

Once they arrived at the customer site, Kevin began to pull out the equipment and talk through how to calibrate a steam flow meter.

"Nope." Adam stopped him. "You are going be my supervisor. I'm going to take our calibration procedure and walk through the evolution from start to finish."

"But, Adam, some of the info in that procedure is dated, and you have to know . . ."

"Stop. This is what we are going to do and then we'll talk about it afterward."

Standing in front of a steam vortex flow meter, Adam reviewed the procedure. The first step identified the equipment required. Adam went down the list one at a time: Rosemount meter.

"Okay, Kevin. Which of these is the Rosemount meter?"

"Well, we don't use a Rosemount meter anymore. We switched to the Yugaski."

"When?"

"About 10 years ago"

The evolution didn't get any better. The procedure was completely unusable. Every step required 15 minutes of explanation from Kevin or was completely irrelevant based on new technology of testing devices and the meter itself. The procedure was a relic and belonged in a museum not a technician's equipment bag.

The real kicker was that the reason the meter was reading incorrectly was because the installation settings (like the series of questions a new printer asks before its first use—language, time zone, printing preferences, etc.) had not been modified from the factory defaults. Of the 13 inputs, seven were incorrect.

"Kevin, do we have a procedure or any written guidance on how to setup a meter."

"No. But it's not that complicated; any professional technician should be able to figure that out without a procedure."

"Kevin, 'should' is the most dangerous word in our profession. Further, since Josh has been working here for seven years, and his professional competency has never been called into question, I think your thought process is flawed."

But that's not the knock out punch. Get ready for it . . . . the calibration procedure never addressed checking the meter inputs. Therefore, even if the calibration was updated and used, the mistake made during installation would not have been caught during the calibration process.

Adam's mind was spinning with thoughts. *The fact that the techs were carrying around a procedure that the supervisors knew was antiquated meant that we have created a hypocritical environment that the technicians were placed into unfairly. If the supervisors were still half-heartedly preaching procedural compliance to the team, the problem was with our culture and the supervisors. Further, if we truly had a culture that respected procedures, Josh or another technician would have clearly raised his hand and pointed out that the procedure needed to be updated.*

This example serves as an illustration that many times things are not as they appear. The truth is often buried in a sea of questions about standard procedures that require a leader to accept answers that may be uncomfortable and unsettling. However, the alternative is to turn a blind eye to elements of our organization that require attention. These efforts build upon themselves and can incrementally transform an organization from mediocre to world class.

ENFORCING STANDARDS

Standards are the front and back door of our organizational structure. If you want to be a member of the organization, standards represent the uncompromising expectations that you must meet. If you prove that you can't or won't, the organization will show you the exit door.

If actions of the organization are not enforced with this "in or out" philosophy, then its standards aren't really standards. They are polite requests. Creating a list of strong words like *Integrity*, *Professionalism*, *Safety*, and *Customer Service* and then placing them on fancy posters in a break room is easy. Just because that poster claims that these words represent the company standards doesn't make it so.

When it comes to standards, expectations are a good starting point, but ultimately the organization will receive what its leaders inspect and tolerate, not what it expects. This is significant when identifying what your organizational standards are because you must be willing to enforce them, at all levels in the chain of command. If you are not, you will create a hypocritical environment which is, arguably, worse than not having established the standard at all.

Maintaining standards is no simple task. Often, when a standard is established there is a lot of emphasis on the standard. The leaders of the organization will (and should) go to great lengths to clearly define the standard, communicate the standard, the rationale behind the standard, and the organizational benefits to adhering to the standard. Immediately following this communication, there are typically actions taken by the organization's leaders to validate that the standard is being adhered to; this may include actions such as additional periodic training sessions, written quizzes, and inspections.

Regardless of the methods employed, standards will never set themselves. Even with the most talented and motivated workforce, if an organization's leaders are not taking active steps to evaluate performance and trends of performance, standards will remain standards in words only. Standards translate to actions and subsequently contribute to the organization's operational excellence only when the leaders are willing to see and accept the truth. Leaders will never be able to observe or learn the realities of their organization's standards of operations from the comfort of their offices or their conference rooms. The truth is found where work occurs, and until the leaders insert themselves periodically into that work environment, the gap between what is expected and what is will remain and is likely to grow over time.

CHAPTER 4

QUESTIONING ATTITUDE

To the uninitiated the principle of a Questioning Attitude may sound like it is an anti-authority sentiment ripe with suspicion and doubt. Perhaps, you imagine an environment where an organization's members are constantly questioning the actions and motives of their peers and the direction and orders of their superiors. That doesn't sound healthy and, fortunately, that is not what we are discussing.

WHAT IS A QUESTIONING ATTITUDE?

"Questioning Attitude" is the third principle in the Navy's Nuclear Submarine culture. It is a thought process that employs a person's fundamental knowledge to critically evaluate the processes, ideas, or operations that her organization uses. An effective critical evaluation requires an analysis with a fresh set of eyes in order to be set free from the habits that blind us most. This fresh look is most effective when based on fundamental knowledge or "first principles."

"First principles" are the principles that pull our thought process away from the cauldron of uncertainty and complexity that stem from the mixture of variables, opinions, and habits and towards the refreshing simplicity of knowns and basic principles. The use of first principles can be a figurative (but sometimes literal) timeout utilized when members of an organization need to make a decision or find a solution to an issue in the face of a mountain of data, information, and opinions that are often at odds with each other. Resorting to first principles, allows the thought process to retrace its steps back to the fundamental truths of the issue. This often (usually) brings the thought process back to well-understood axioms or principles such as Newton's laws of motion or supply-and-demand curves.

This allows a group of people to answer the question, "What are we sure is true?" If we can answer that question, then we can move one deliberate step at a time until we reach the solution. Of course, this is easier said than done—especially when the first principles methodology is not yet embedded into the culture.

Consider the following thought experiment described as "Monty Hall's Dilemma."

You are a contestant on the famous game show, *Let's Make a Deal*. The host, who at the time of writing of this book is Wayne Brady (but was originally Monty Hall), shows you three rooms. Each is behind a closed door, and you cannot see what is inside; however, you know that in one of the rooms is an expensive prize (a vacation, a car, etc.) and the other two rooms have worthless prizes (goat hay, old newspapers, etc.). Wayne Brady asks you to a pick a room that that you want. You select Room Number One.

Wayne Brady then opens the door to one of the rooms that you didn't select and it contains the loser prize—a pile of empty Pepsi cans. Now there are two unrevealed rooms remaining— one has a valuable prize in it and the other does not. You are currently holding onto Room Number One—your original selection.

Brady asks if you would like to change your choice and switch to Room Number Three (the other unrevealed room). What should you do? Should you switch, or should you stick with your original choice? Or does it matter?

Let's start our discussion with the answer. The probability of winning goes up substantially when you switch. So let's assume that this represents a process within an organization. To most people, there are no intuitive reasons to switch, and it is not difficult to imagine people (and you may be one right now) who are willing to defend this process of *both doors have the same probability of winning*. "That's the way we've always done it!"

But when we use first principles with a questioning attitude, a different solution is revealed. **HERE**

1. What is the probability that we picked the correct door the first time? The answer is 33% or 1/3. Is there anything about that percentage that we doubt? If we picked a random door one million times, do we expect to be correct approximately 333,333 times? Yes. Are we sure? Yes. So my original selection has a probability of being correct 1/3 of the time. Is there anything that anyone could do (outside of insider information) that can change that probability once the decision is made? No. Okay, so we all agree that our chances of being correct with this methodology (random selection) is 1/3.

2. If our door has 1/3 chance of being correct, then the combine odds of the remaining doors must have a 2/3 chance. Right? The total odds of the three doors must equal 1.

These two facts are known and undisputable. They use basic statistics and steer clear of intuition, opinions, or complex explanations. The basics: You have 33 percent chance of being correct and the other two doors, combined, have a 67 percent chance of being correct. Let's state that last fact mathematically A + B = 67 percent. Wayne shows you A and you know now that A has a 0 percent chance. Therefore, the remaining door, B, has a 67 percent chance of winning. By switching you increase your odds from 33 percent to 67 percent. This is a fact and you can argue it all day long, but first principles lead us to the correct answer even if that correct answer is not intuitive.

Most of our organizational concerns aren't game show gimmicks, but this example serves as a reminder that sometimes even our strongest instincts are incorrect. Sometimes the way we have been doing a particular process can be improved. If the improvement was obvious it would have been made already. This is why we need our Nuclear Submariners to also be thinking "what if", "why", "what about this", "what about that" - this is a thought process that when combined with first principles often leads to unexpected results that can improve an organization's performance by identifying options or improvements that were not obvious or intuitive.

Perfection is unattainable. No process, operation, or evolution will ever be perfect. Therefore, there is always room for improvement. An organization's performance can always be improved. How? Many organizations rely on a separate group that often go by names such as "Operational Excellence Group" or "Organizational Excellence Group." There is no room for such a group on board a submarine, instead a Nuclear Submarine relies on instilling a questioning attitude into each and every member of the crew. The innovation ideas and process improvements are a byproduct of this attitude.

More simply, the questioning attitude that we are discussing is an attitude that rarely accepts something at face value. The right questions at the right time to the right person can most always reveal an unexpected truth - sometimes a welcomed one, sometimes an unwelcomed one, but if "truth" is the objective to which we hold our allegiance, we learn that it is almost always buried deeply behind a series of questions. Recalling that our goal is a culture of operational excellence, the cumulative impact of each member of the organization learning to think critically about what he is doing at any given time provides massive returns on the investment spent in developing those critical thinking skills.

We have emphasized that a questioning attitude is a THOUGHT process. The next chapter will discuss how those thoughts are communicated most effectively in the organization. This point is important because there are many times on a submarine where immediate and decisive action is required. This requirement is a necessity to save lives. The Commanding Officer (or his representative, the Officer of the Deck) often have mere milliseconds to make life and death decisions. Therefore, when the Officer of the Deck, orders "Left Full Rudder" - the helmsman is expected to immediately repeat that order ("Left Full Rudder, aye") and then immediately bring the rudder to full left. That is not to suggest that his mind shouldn't be actively engaged. "Why did the Officer of the Deck order that?" "Why left instead of right?" "Why left full instead of left hard rudder?" He absolutely should be asking himself these questions. This is an engaged mind and we train and expect it from all crew members.

Why is a Questioning Attitude a Principle of Excellence?

Let's back up a moment and discuss why a culture that possesses a questioning attitude is likely to excel when others falter.

Consider your commute to work. In all likelihood you follow a pattern and routine that is so well embedded in your brain, that you hardly even think about it. You just start the car and the next thing you know, you are at work. However, are you sure it is the most efficient route to work? How do you know that? Have you identified the alternatives? If so, how long ago did you do that? Have the alternatives changed since then? If you haven't thought about this very basic question in detail in the last three years, there is a greater than fifty percent chance there is a change that you could make that would save fuel and/or time. How do we know this? Because we have witnessed the process of examining alternatives that have been heralded as dead ends, only to find — over, and over, and over, again, that the persistent and correct line of questioning is successful in over fifty percent of the time - conservatively.

Incremental change matters. Incremental progress is always available and if you are not looking for it, we can assure you that a competitor of yours is. Again, many organizations seek to create "Operational Excellence" groups. We know these groups have their place and value, but that should be above and beyond what can be extracted from the cumulative effort of each organization member.

In the following story, consider the costs when questioning attitude is not ingrained as a cultural principle.

Sherry was the General Manager of a food manufacturing plant. As a cost savings effort, Sherry collaborated with their chemical provider and chose to switch one of their chemicals. The purpose of the chemical was to scavenge oxygen in their water source to reduce corrosion in the plant and their products. The new chemical was a sulfide based product.

Sherry and the plant manager, Mike, discussed the transition and it seemed pretty simple, if not uneventful. No operational changes were required and the laboratory technicians had been briefed and trained on the required sampling techniques for the new chemical. They agreed that when the existing inventory of chemicals were depleted the new chemical would be added. Easy as pie.

The new sulfide based chemical was being warehoused until the plant needed. Their chemical supply representative, Rick, assured them that the new chemical could be delivered within 24 hours. After a few days of depleting the existing chemical, the on-shift plant supervisor made the call to request delivery of the new chemical. When the chemical truck arrived, he supervised the workers pumping it to their chemical storage tank. As he watched the chemical flow from the truck to their tank, Rick noticed something unusual about the new chemical. It had what appeared to be crystallized particles in it. He shrugged it off and assumed it was expected - what did he know about sulfide oxygen scavengers?

The chemical tank was stored in a warehouse but was not climate controlled. The temperatures in the depths of February had been below freezing for weeks. This error caused the sulfide to come out of solution (not expected and not good). White wine has this same phenomenon because it contains sulfides, you will sometimes see crystallized particles on the bottom of a bottle of white wine if it has been in subzero temperatures.

The crystallized particles immediately clogged up the chemical injection pump. This meant that not only was an injection pump potentially damaged but no chemicals were being added to the plant. The first supervisor to notice a problem with the pump, ordered the pump to be replaced. No further action was taken to investigate why the pump failed.

The laboratory technician, Gary, was sampling as instructed. His results revealed elevated oxygen levels. He tested oxygen levels every day for years and result was always zero - always. He never saw numbers or trends like he was now seeing - 50 parts per million (ppm), 75 ppm, 100 ppm. He assumed that he was either sampling incorrectly or the new sampling equipment was faulty. He took no immediate action, but rather made a note to talk to the chemical provider representative the next time he saw him.

The auxiliary operator logs, among hundreds of other plant parameters, the level of the chemical tank. Because the injection pump was clogged (yes, the second one was gunked up as well), no chemicals were being injected into the plant, therefore, the tank level remained constant. The auxiliary operator was accustomed to seeing an inch or more reduction in tank level each day. He assumed that the new chemical was used less. He thought nothing more of the constant level and went about his way.

The situation got better before it got worse, but for the purposes of this chapter, we'll stop there. From management to the operators, there was no instilled sense to ask "what if?" and "why?" as a part of their standard thought process. When the dust settled, the plant suffered nearly $1M in damages.

This is real money and it's based on a real life story that probably happens more than we all want to admit. If establishing a questioning attitude is not instilled in the members of an organization with deliberate and persistent effort, not only will the organization not reap the benefits of innovation and creativity, it will also fail to catch abnormal situations while they are still manageable.

Consider the following questions and the impact they could have had if they were thought of and then acted upon.

Management: (Shelly and Mike)

> What could go wrong?
> What's the worst that could happen?
> What if the chemical doesn't work?
> Do we need to have a training session with supervisors about this new chemical?
> Has any other plant tried this?

Supervisor that oversaw the chemical delivery:

> What is that "stuff" that I am seeing?
> Should I let someone know?
> Should I ask the provider if those particles are expected?

Supervisor that ordered the pump replacement:

> Why did the pump fail?
> Should we examine its internals before ordering a wholesale swap?
> What causes pumps to fail like this? Have we seen similar failures in the past?
> Did anything in the chemical system recently change?
> After the replacement, should we monitor the pump more closely?

Laboratory Technicians:

> Why is oxygen elevated?
> Why do we sample for oxygen?
> What could happen to the plant if these oxygen levels

are accurate?

Is there a backup sample technique that I could use to check these results?

Do we have evidence to support that there are no chemicals being added to the plant?

If there were no chemicals being added to the plant, what would oxygen levels be?

Auxiliary Operator that logged tank level:

Why did the tank level remain constant over a 24-hour period?

What could cause this?

Is this an expected indication or trend?

Should I report this?

Anyone of these questions could have averted the damage that subsequently occurred. In most cases, these questions do cross people's minds but only as fleeting thoughts. The fact that these fleeting thoughts aren't captured, evaluated, and dissected is reflective of a lack of something. This deficit may vary from person to person, but they are typically lack of ownership, lack of training, lack of conscientiousness, or lack of communication. Each of these are worth exploring:

Lack of ownership: Do the members of your organization treat their duties as if the buck stops with them? If this scenario involved the people's homes or cars, would the individuals have taken more aggressive and effective action? The challenge with establishing ownership in an organization is that members can rationalize their own inaction by convincing themselves that someone else will deal with problems. Often this stems from previous perceived slights from the organization. For example, a member has an idea to increase productivity and his boss responds with, "That's not your job to worry about. Just stick to what we pay you to do." If you receive this message from the organization, directly or implied, disengaging your brain from the organization's objectives is a logical conclusion. This may read like an exaggerated message. *What kind of supervisor would say that?* The answer is simple - real ones and it happens everyday whether they use such direct language or just ignore a person's ideas or questions.

Lack of training: The expectation of a questioning attitude requires training. There is no shortcut around this. An effective training program that drills the expectation of evaluating and questioning the information they process during the course of their day is required for even the most talented and willing employees and supervisors. Further, establishing effective and consistent questioning attitudes in your organization is not a destination and it is not an on/off light switch. It is more like holding a spring in place, as soon as you let go the spring snaps back. The reason for this is that we are attempting to develop a way of thinking that is against our nature. The human brain seeks the comfort of patterns and habits whereas an effective questioning attitude is always

Lack of conscientiousness. Sometimes a person simply fails to do their duty. The expectation of ownership has been clearly defined for them, they understand and exhibit questioning attitude thought process, but they simply fail to follow through with them. They allow their fleeting thoughts ("Shouldn't oxygen be zero?") to be overshadowed by laziness or apathy. If this deficiency is observed, it is tempting to blame that one individual. Be careful of doing so because there may be an underpinning cultural issue that is impacting the entire organization.

Lack of communication: Sometimes we know what we don't know — such as, "I don't know the recipe for that meal." Other times we don't know what we don't know. In the example about the oxygen scavenger chemical switch, the auxiliary operator may not have known that sulfides crystallize in cold temperatures; worse, he doesn't know that he doesn't know. As we gain years of experience, we learn more — hopefully. Therefore, it is sage advice to never be the "senior person with a secret."

This could be more plainly stated as it is critical to keep your boss informed about all important matters. What rises to the level of "important" is the catch. Therefore, when we observe something that is different and we don't understand much about it, other than, "this looks different," we must remind ourselves that we cannot know what we don't know. However, our boss might. As soon as you choose to be the "senior person with a secret," you are also choosing to retain the responsibility for the consequences of that secret. This is unwise, if not foolish, when that secret is one that you don't quite understand.

River Water Project

Consider the story of Adam, a new General Manager, at a power plant located directly on the banks of one of the nation's largest rivers. The power plant was rife with problems - financially, operationally, and culturally. The plant was owned by a sizeable international company that was committed to resurrecting the operation but wasn't certain how to do so. Adding a General Manager position on site wasn't the company's first major effort to turn the plant around, but they hoped it would be the last one.

As Adam was acclimating himself to the plant and its team of operators and supervisors, he was reviewing the engineering studies that lined one of the shelves of the bookcase in his office. These explorative studies spanned several decades and each was different than the next, but they each attempted to answer the question: "Was it financially feasible to use river water for steam production?" (The plant was a district energy plant that provided its customers steam for heating or industrial use through miles of underground piping)

The plants' water and sewer costs represented approximately one-fifth of their variable costs and the costs of both had been rising aggressively over the past few years. That trend was scheduled to continue as the city publicly announced their intention to double sewer costs overs the next ten years. This rise was eating into a margin that was already struggling to remain black. Any water or sewer savings would go right to the bottom line. Adam read the reports with a sense of hope, but each executive summary after the next pronounced the study's conclusion — "Not cost effective." One by one, he placed the studies back on the shelf where they could return to collecting dust.

Adam was perplexed. *The Egyptians were able to purify the Nile River through the principles of coagulation in 1500 BC, but we can't purify river water at our doorstep?*

The company had a talented matrix of professionals that were always available to assist - engineering, finance, accounting, environmental professionals all extraordinarily capable and eager to assist Adam in all aspects of plant operations. However, when we began to probe them about the possibility of using river water for steam production, the answers were all the same.

Finance: "We've already looked at that a million times. We can't make the numbers work."

Engineering: "The river's levels of silica, turbidity, and alkalinity are too high. Won't work."

Environmental: "Once you filter the water, the state won't allow you to return the reject water back to the river."

Accounting: "Even if you could make the numbers work, you don't have a capital expenditure envelope large enough for a project this size."

Operations: "If you improve the quality of the water, surface blow-down percentage will go down, and we use that blowdown steam to pre-heat the feedwater. You might actually cause efficiency of the plant to go down."

Janitorial: "My mops won't be able to withstand the extra leakage that system would create."

The janitorial comment was facetious but representative of the "no, no, no" gauntlet that Adam was encountering. Sometimes the answer truly is "no" and Adam knew this, but given the potentially game-changing cost savings, he was inclined to be a bit more stubborn than might be welcomed of a new General Manager. The number of "no's" were a bit overwhelming, but the rationale for each, by itself, seemed paper thin.

Adam chose to tackle the engineering concerns first. If he could identify a new way of engineering a solution, perhaps the other aspects would fall into place.

He scratched, clawed, and climbed over the tall engineering wall of "No" - one question at a time:

"Why is the RO (Reverse Osmosis) unit sized the way it is?"

"Why are we making this quality of water? Where did those specifications come from?"

"Is there any other technology being used?"

"How often does the river water contain the high levels of organics that the unit is sized for?"

"Would the state allow us to discharge the sludge back to the river if contains no chemicals?"

"Is there a chemical-free solution?"

"What does the project look like financially if we make city quality water instead of boiler quality water?"

"What does the project look like financially if we mix city water with a smaller filtration unit?"

The journey was a year of slowly removing each brick of the "No" wall, until finally a solution emerged which was a smaller and simpler design that created city quality water 80 percent of the time (to account for seasonal extremes in the river water chemistry). The dust settled, the project was approved, and the company enjoyed a water and sewer savings of nearly two million dollars a year. Sometimes, we fight with questions. Questions, like a left jab, improve with practice and persistence.

Establishing a Questioning Attitude Culture

When an unexpected, unintended, or undesirable outcome occurs on a Nuclear Submarine, the event is evaluated in detail through a process known as a "critique" which is described in Chapter Seven. Without exception, a contributing factor is usually that one or more people involved failed to demonstrate an effective questioning attitude. This is true despite the immense amount of emphasis and training that is devoted to developing a questioning attitude in all crew members. If the Nuclear Submarine community continues to deal with deficiencies derived from the lack of a questioning attitude, how can an organization that doesn't have the benefit of decades of cultural strength possibly develop this way of thinking in its members?

It's a good question without a simple answer but we'll introduce some concepts that can help.

Train Supervisors First

If you find that your organization lacks the questioning attitude culture, your chances of instilling this cultural characteristic are increased by training supervisors first. There are several reasons for this. The first is that anytime you have a new concept to introduce, introduce it with supervisors first because supervisors are more likely to give you the feedback that you need. They are more likely to challenge the message and the manner in which you are delivering it. If you can work through this process with the supervisors, not only do you achieve their buy-in first, but you will be able to identify how better to communicate it to the rest of the organization. Also, you now have multiplied the number of people capable of communicating it to the rest of the organization.

If you begin the learning process with supervisors, you have a better chance of minimizing the possibility that you create a "runaway questioning attitude train." Like many powerful weapons, if the questioning attitude principle is not handled with care it can be destructive. Specifically, you can inadvertently create "friendly fire" casualties when everything that everyone says and does is questioned. That is the runaway train that we want you to avoid. Unfortunately, this phenomenon is not uncommon. Even in the Nuclear Submarine culture there are commands whose questioning attitude gets the best of them and the net effect is a team that grows suspicious, distrustful, and ineffective. There is a delicate balance that an organization must seek when developing the questioning attitude culture and this can and must be emphasized to the supervisors. Taken to its extremes and without proper supervision, you'll have an organization that begins to question each other on everything that they do along with an abundance of unnecessary critiques (Chapter 7).

Disrupt Thought Patterns

A successful questioning attitude begins with the ability to see the work environment with a fresh set of eyes. There is no more powerful skill than to look at a situation for what it is and not what we think it is or what we expect it to be. Unfortunately, our brains are obsessed with developing patterns. Once a particular object or process is categorized by our brains - not only by identifying it, but by assuming what is going to happen and how. It is precisely that assumption that we want to get at, because when the brain makes assumptions about outcomes, it is very frequently wrong. This isn't due to lack of information, in fact the opposite is true - the brain processes too much information; this is due to our brain's incessant need to immediately identify what is going to happen next - second to second, millisecond to millisecond.

Consider a baseball player hitting a 90 miles per hour fastball. Neuroscience concludes that it should be impossible to hit a 90 miles per hour fastball. At that speed, the ball reaches the batter in 400msec. Consider that it takes the brain 100 msec to process an image, 25 msec to signal the body to swing, and 150 msec to swing. This means that a baseball player has only

125 msec to evaluate whether to swing or not. By comparison, a blink lasts 300 msec. How could a person possibly make this decision? One Yale physicist, Robert Adair[11], concluded that it is impossible to do this based on what he know of the human brain. The only reasonable explanation provided is the power of our brain's imagination. The brain of a baseball player is able to take an iota of data and assign a likely future outcome (speed, location) and this process occurs at a layer beneath the player's consciousness which highlights the always underestimated brilliance of Yogi Berra who once said, "You can't think and hit at the same time."

What's the connection between this and developing a questioning attitude? The connection is that it takes deliberate, thoughtful, and repetitive efforts to overcome our brain's propensity to see what what it wants to see. It takes a conscious effort to evaluate the data in our environment objectively and to identify the details that our brains may be ignoring because of its tendency to draw lightning fast conclusions on a limited amount information. The feat of hitting a baseball squarely is a remarkable one but even the best in the world have a 30 percent success rate. The brain operates this way to maximize our chances of surviving a bear attack that originates in your peripheral vision not to objectively evaluate the process and procedures of an organization's business model.

Did you ever place something in a location in your home temporarily only to find that it remained there for four months? Chances are that the object looked out of place the next morning, but then something peculiar happens. Your brain rapidly concludes that "it has always been there" and the next morning, it's existence doesn't even make the cut to your conscious thoughts. The same is true in our professional lives. One day

1   Adair, R. (2001). *The Physics of Baseball* AR: Harper Perennial

the boss says, "Can everyone please place their expense reports into the folder labeled 'Trips to the zoo' on the shared folder?" Of course this seems laughably odd, but it wouldn't be surprising if three years later and three bosses later, expense reports were still being filed there.

A Skill, Not a Decision

An effective questioning attitude is a skill developed over time not a decision by leaders that an organization should just have it. That may be the most important sentence in the chapter. An effective questioning attitude is a skill not a decision. It is not uncommon to see an organization stumble because it fails to recognize this. Consider the following example:

A Nuclear Submarine has just returned from a six-month deployment and is enjoying a much deserved one-month long "stand down" period. This means no maintenance, training, or missions. The submarine, of course, is still manned by a skeleton crew each day called the duty section. The reactor is shut down and the reactor systems are run by electrical power provided from shore connections — aptly called "shore power." This shore power is backed up the ship's electrical storage battery and emergency diesel generator. These backup sources can provide power for an extended duration but we much prefer them sitting idly as backup power rather than using them. Therefore, if shore power is lost, the sense of urgency to restore shore power is high, as is the necessity to determine what caused the loss of shore power.

The second day of stand down is kicked off at 0500 (5 a.m.) with a 1MC (the ship's main public announcement system). "Loss of shore power. Prepare to snorkel." ("Prepare to snorkel" is an order to make the emergency diesel generator ready for use.) The ship finds the shore power breakers tripped and is able to reset and shut them in short order. The duty section determines shore power tripped off-line because of the large starting current of the low pressure blower which is used daily to remove any water from the ship's main ballast tanks — keeping these tanks full of air instead of water is what allows the submarine to float rather than sink at the pier. The Commanding Officer grills the Ship's Duty Officer about the assessment of the cause of the loss of shore power. The starting current of the low pressure blower is high, but if the Electric Plant is operated properly it should not result in a loss of shore power. Although, not convinced, the Commanding Officer lets the topic go because shore power was restored quickly and there was no other information available to pursue.

The very next day, the same events occur and the same reason is presented. This is now another duty section (different group of people) and the Commanding Officer calls the Engineer and tells him to figure out what is going on. *Did two separate duty sections both mismanage the ship's Electric Plant in the same way when starting the low pressure blower?* This seems unlikely given that the low pressure blower is started every day while in port to remove any water build up in the main ballast tanks. The Engineer is unable to uncover anything but comes to the submarine the following morning to observe the low pressure blower start. The Engineer observes the evolution from the Engine Room's Control Room (called the Maneuvering Area). Shore power is lost . . . again. After restoring shore power, the Engineer, following a hunch, has the duty section start the low pressure blower but this time, he directs the Electric Plant Operator to limit the amount of the low pressure blower's starting surge that is pulled from shore power to 3/4s of the shore power limit. No loss of shore power. He directs the evolution again using the same guidance. No loss of shore power. He directs the evolution a third time, but this time directs them to limit the amount of the low pressure blower's starting surge that is pulled from shore power to just below the shore power limit — which is the typical manner of conducting the evolution because it minimizes the use of the ship's electrical storage battery. Shore power is lost again. The Engineer's hunch was correct. Shore power is connected to the ship through four breakers, one of the breakers must be faulty and the ship was only receiving 3/4s of the usual amount of power from the shore connection.

The Engineer was able to deduce what was occurring because he had the benefit of more experience than the other supervisors. The other supervisors failed to exhibit an effective questioning attitude, because they did not know which questions to ask. They lacked the skill, not the willingness to do so. However, they should have been able to recognize this themselves because the answer they accepted was not supported by facts. The ship starts the low pressure blower every morning the ship is in port. If their conclusion was reasonable, the electric plant was operated incorrectly. If the electric plant was operated incorrectly, where was the corrective action or lessons learned from this mistake?

The Commanding Officer convened a meeting with the the supervisors from the previous two duty sections to discuss the situation with them. During this meeting, the Engineer reviewed the thought process that the he used to determine what actually resulted in the successive loss of shore power. He further discussed why the explanation that they each accepted was not well-thought out or logical. The Commanding Officer then discussed his expectations should they find themselves in a similar scenario in the future. He acknowledged the difficulty of dealing with a situation where the answer out of reach because "you don't know, what you don't know." However, he further explained the dangers of accepting an answer that cannot be supported with facts. He discussed the utilization of solving problem tools such as:

1. Asking supervisors for help.

None of the supervisors specifically asked the Engineer or the Commanding Officer for assistance in determining the cause. Although, supervisors on a Nuclear Submarine are generally discouraged from asking their supervisors to figure out their problems for them. Leaders are always willing to help. So while "I don't know what happen. Do you?" would not have be received well. "I am having a difficult time determining how to proceed in identifying possible causes, can you assist me?"

2. Resorting to first principles. What did they know? The shore power breaker tripped.

What else did they know? They knew the list of conditions that can cause the shore power breaker to trip. They could have listed each of these trip conditions and then evaluated each. Further, they could have listed each component between the shore power breaker that tripped open and shore power itself. That list is manageable in length and could have been evaluated through a process of elimination.

3. Involve the subject matter experts.

Although the Electrical Division Chief was consulted, the supervisors had other experts at their disposal that were not consulted such as the Squadron Electrician, the Engineering Department Master Chief (the senior enlisted person in the Engineering Department), and Petty Officer John Smith (the electrician that has been supervising the procedure for hooking up shore power for over a year).

The corrective action for these supervisors was to prepare a training session for their duty sections that addressed the events that occurred and the steps that could have been taken to derive to a more technically sound solution. This post-event process (the meeting with the Commanding Officer and their training preparation) all served to increase the supervisor's skills in handling similar situations in the future. No one in the chain of command accused the supervisors of not wanting to have an effective questioning. The situation was addressed with the intent of improving the supervisor's skill set strength in exhibiting an effective question attitude going forward. Beware of failing to acknowledge that an effective question attitude is a skill not a decision. Specifically, if the Commanding Officer had simply admonished the supervisors for failing to have a questioning attitude, he would have missed the opportunity to increase the strength of his talent pool of his command.

Start Small

Rome wasn't built in a day — neither was any other city in the history of the world, but that's not the point. Given the fact that an effective questioning attitude is a skill and not a decision, we should train our people consistent to this fact. We wouldn't start a novice guitar student with a Jimi Hendrix guitar solo. But when the stakes are high, such as they are in an organization's operational and financial performance, it may be challenging to accept the "start small" methodology, but we must.

Earlier in the chapter, we discussed the probability that if you evaluated your morning commute carefully enough, you would probably uncover a change that you could make to shave a minute or a mile off of your route. Consider an exercise such as this one with the group. Have them evaluate a process that is not work related and force them to think through every detail, all the while asking questions like "What if?", "Why Not?" For example, are they sure that public transportation wouldn't be cheaper or faster. Have they evaluated Uber or Taxicab by considering the cost of gas, car maintenance, and even work that might be able to do during the ride. Using scenarios from people's lives is a good training topic because you can be assured that they are immediately interested . . .it's their lives. You could expand the topic to their process for yard or house work — have they considered all options and asked as many questions as they can even if they "think" they know the answer already. Resort to first principles and run every possible option into the ground. This methodology may sound familiar to the reader that has studied Lean and/or Six Sigma process control. These are examples of programmatic questioning attitude processes to improve the operational performance and reliability.

Leverage Fascination

Ultimately, you will want to shift to items with operational impacts not people's commuting costs and times. Again, we recommend to keep these topics off the beaten path and fascinating. What's fascinating? Consider the following example.

While at sea, operations in the Engine Room of a Nuclear Submarine can be mundane at times. It is not uncommon for a submarine to operate at a very slow speeds for extended periods of time while conducting missions of that require great stealth. As a result, Engine Room operations can be limited to making steady-state steam for propulsion and electricity and being quiet. The latter often precludes performing even routine maintenance because the risk of making metallic noise is too great.

During these less active times it is incumbent upon the Engineering Officer of the Watch (EOOW) (the officer that supervises all Engine Room activity) to conduct training with the control room area watchstanders during the six-hour long watch. Often the topic and methodology is left to devices and imagination of the Engineer. The challenge of getting Nuclear Submarine operators excited about training three months into a seven deployment cannot be overstated. What works and engages the operators is when the topics are fascinating. For example, assume the EOOW decided that he wanted to train on atmosphere control equipment. The standard (boring) way is to review with and then quiz the operators on the various equipment specifications and operations. Example:

EOOW: Okay, let's review some atmosphere control equipment. Reactor Operator draw a one-line diagram of the Oxygen Generating system. Electric Operator draw a basic block diagram of the control circuitry for the scrubbers (Carbon Dioxide removal system), and Throttleman (Steam Plant Operator), you sketch the location of the Engine Room fans.

Thirty minutes passes and then the EOOW collects the work of each of the operators and discusses any errors or points he considers interesting with the group.

Contrast that with:

EOOW: Reactor Operator, if we isolated the Engine Room from the forward compartment, how long would it take for us to die. That means no oxygen generation, no scrubbers running, and the Engine Room completely isolated from the rest of the ship, how long would we have to live?

Reactor Operator: (Laughs) I have no idea. Why would you ask that, sir?

EOOW: Because I think that's something we should know or at least be able to figure out. What information would we need to answer this question?

Electric Operator: The only oxygen left would be what was in the air and the aft (Engine Room) oxygen banks (tanks full of oxygen) but I don't know how long that would last.

EOOW: Okay, that is a start. We would need to know how much oxygen is stored in the oxygen banks and how much oxygen is in the air. What else would we need to know?

Throttleman: The number of people in the Engine Room. We know there are fifteen people in the Engine Room.

EOOW: Okay, what else?

Reactor Operator: How much oxygen does each person need per hour? Damn, I used to know that number, but I forget.

Throttleman: (Paging through a reference book) I found it. The average person consumes 15 cubic feet of air per hour.

Reactor Operator: Damn - that's right. That sounds familiar.

EOOW: Why don't we make this simpler . . . but much less desirable — assume that there is no oxygen in oxygen banks. What information would we need to figure out how long we have to survive?

Electric Operator: We would need to know how many cubic feet of air is in the Engine Room.

Throttleman: Oh, I know that. It's in this chapter.

EOOW: Wait - stop. Put the book away. Could you figure out without looking it up?

Throttleman: (Laughs) Yeah, if I could remember any of my 10th grade geometry, but I'm not so sure I could. If you ever meet a woman named, Mrs. Niznik — please don't tell her I forgot how.

(Laughter)

EOOW: You can figure out the volume of a rectangle though, right?

Electrical Operator: Length times width times height?

EOOW: Okay so what size rectangle would the boat fit into?

Reactor Operator: 300 feet times 30 feet times 30 feet. If you lop off the sail.

Throttleman: (pounding on calculator) That's 270,000 cubic feet and if we assume the Engine Room is half of that, we get 135,000 cubic feet.

EOOW: Should we take a percentage of that since the submarine is more cylindrical than rectangular?

Throttleman: (on calculator still). Okay, maybe 75 percent? That would be 101,250 cubic feet.

EOOW: Okay. Let's call it 100,000 cubic feet. Now what? What do we know so far?

Reactor Operator: We know that we have 15 people breathing 15 cubic feet of oxygen per hour and 100,000 cubic feet of air.

EOOW: Now what? We want to know how long until we all die.

Electrical Operator: With 15 people breathing 15 cubic feet of air per hour that's 225 cubic feet of air per hour that is breathed in and 100,000 cubic feet of air to breathe. So 100,000 divided by 225 is 444 hours.

Reactor Operator: No, stop. Something isn't right. My recollection is that we would die from carbon dioxide buildup, not from oxygen deprivation. I forget how that works. Engineering Officer of the Watch, can we ask Doc (the Navy Corpsman assigned to the Nuclear Submarine) to join us?

Moments later the ship's "Doc" arrives and they bring him up to speed.

Doc: Wow, you guys have quite the science experiment going on back here. You are right, you guys would die of carbon dioxide poisoning, not oxygen deprivation. Generally, we consider an environment that has 10 percent carbon dioxide to be deadly.

Throttleman: So what's the concentration of carbon dioxide in the air now? The air we are breathing?

Doc: With the fans running and the scrubbers on the concentration is near zero. The fifteen people would be creating the carbon dioxide.

Electrical Operator: So we need to calculate how long it would take us, the fifteen people, to generate 10,000 cubic feet of carbon dioxide — ten percent of the 100,000 cubic foot volume. Right?

Reactor Operator: That sounds right. Doc, I seem to remember that the average person generates about 1 cubic foot of carbon dioxide an hour. Is that right?

Doc: It's right, if you all stay calm. But you'd probably be panicking and that would drive your generation closer to 2 cubic foot per hour.

Reactor Operator: Okay, so that's 2 cubic foot per hour and with 15 people that's 30 cubic feet per hour.

Throttleman: At 30 cubic feet per hour, it would take us 333 hours to die of carbon dioxide poisoning, or two weeks.

Doc: That sounds about right, but you guys are being selfish. Unless the forward guys are trying to murder you, wouldn't it be more realistic to figure out how long you'd have for the entire boat to perish.

Reactor Operator: That's easy to calculate now. We'd assume 200,000 cubic feet for the entire boat. 120 guys so that's 240 cubic feet of carbon dioxide per hour and we have to calculate how long until ten percent, or 20,000 cubic feet of carbon dioxide is produced.

Throttleman: (his face draws long) That only 83 hours or 3.5 days.

EOOW: See that, Reactor Operator? And you thought it was a silly question. What'd we learn?

Reactor Operator: If we sink, we shut and lock the engine room door.

EOOW: (Laughs) If we ask the right questions and take things one step at a time with first principles we can figure out anything. Good job today, guys.

The Result

We led the chapter with the discussion of why a culture that exhibits a questioning attitude will ultimately outperform one that does not. It is only natural. Human intrigue and fascination is immeasurably powerful but it must be cultivated. There is something about belonging to an organization that tends to turn off people's imagination spigots. It is your job to get these spigots turned on and kept on.

There is another interesting result about living in an organization with a healthy questioning attitude. That is the ability to ask the right questions to get to the heart of a problem even in foreign environments.

When Nuclear Submarine Commanding Officers move on to other endeavors inside and outside of the Navy, they take with them the well-deserved confidence to be able to handle the most challenging of assignments. This confidence is primarily based on their ability to ask the right questions. We expect and train the entire Nuclear Submarine community to maintain a questioning attitude in all that they do, but a successful command tour aboard a Nuclear Submarine is a Naval Officer's Doctorate in developing a keen questioning attitude.

This skill develops out of necessity. It is surprising how many people assume that the Commanding Officer is the man that knows the most about the submarines' systems. This sentiment represents a significant simplification of the complexity of a modern Nuclear Submarine. No one man could possibly be an expert in all of the systems especially since the odds are high that there will be systems installed on the submarine he commands that he has never seen before - upgraded sonar systems, fire control systems, weaponry, communication antennas, and even nuclear reactor plant modifications. He is however fully responsible for every bolt and every soul on his ship. How does he handle this responsibility? By knowing what questions to ask and when. This same level of questioning attitude can create great leaders at all levels of your organization but the culture must be nurtured and intentional.

CHAPTER 5

WATCHTEAM BACKUP

In the previous chapter, we discussed the importance of establishing an organization with a questioning attitude. This concept primarily addresses a way of *thinking*. Watchteam backup brings that concept to the next level as a way of *acting*. The concept is that when all members of the organization are fully engaged in an operation or process, it is likely that someone's engaged mind will identify a point or concept that, if verbalized, could substantially improve the organization's performance. In the most effective and extreme cases, this verbalization can prevent a dangerous or incorrect action from proceeding.

Watchteam backup is the Holy Grail of an organization's climb to operational excellence. The fourth principle of the Nuclear Submarine's culture of excellence, watchteam backup, requires all of the other principles (Knowledge, Procedural Compliance, Questioning Attitude, and Integrity) to operate in harmony. What is watchteam backup? Examples provide more clarity than a definition:

- The team's junior engineer saves the company credibility and money, when she recognizes and then speaks up that the model for a multi-million-dollar project is flawed because the incorrect coefficient of friction is being used.
- A boiler operator is about to ignite a high-pressure boiler out of sequence, but his apprentice stops him

before he does.

- A senior salesperson of medical devices has been performing poorly of late. A junior salesperson, in private, tells the senior salesperson that his presentations have been very low energy lately and she thinks it is impacting his performance.
- A junior financial analyst is in attendance during the CFO's briefing to the CEO. The presentation contains an error that skews the data unfavorably. The junior financial analyst politely shares his observation with the CFO in private after the presentation.
- The frontline supervisor of a manufacturing plant orders the front line supervisor to start #2 Press. A junior electrician overhears this order and knows that #2 Press is disassembled with mechanics working on it. He yells, "NO! STOP!" one second before the Press is started and averts a potential serious injury.

This is watchteam backup.

THE BENEFITS OF WATCHTEAM BACKUP

When employees are engaged in the organization's mission and empowered to communicate thoughts that service the organization's mission, the effectiveness of the organization increases tenfold. The two benefits of this principle having the most impact on an organization are mistake avoidance and organizational effectiveness.

Mistake Avoidance

Every organization is composed of people. People are fallible and, therefore, make mistakes. It is not, however, a natural conclusion that, therefore, the organization will make mistakes.

Effective watchteam backup protects the organization from what we call single-point failures. Leaders must aggressively identify the places in their organization where they are susceptible to single-point failures. Can you develop a process by which the marketing department reviews the engineer's work and vice versa? Can you develop a process by which the administrative assistant is trained in basics of the company's operation? Can you invite a broader, and even junior, audience to participate in your next project proposal? Single-point failures don't exist solely in individuals; they can exist in business units, functional units, and seniority levels. Of course, it's not enough to simply involve these diverse groups—you must invest the time with them to explain to them and sell them on the notion that their opinions are valuable, even if they are not considered subject-matter experts.

Organizational Effectiveness

Although it is challenging to quantify operational effectiveness, when the mechanic learns to keep his ears open about matters that are outside the bounds of his strictly defined duties and responsibilities AND he has been empowered to weigh in on matters outside of his specific lane, he will be the one who has the breakthrough recommendation. He will be the one who asks the question that creates a paradigm shift. Pigeonholing your employees may protect egos, but it fails to leverage the enormous amount of untapped brain power in your organization.

Effective watchteam backup is the golden ring of teamwork. Harry Truman once said, "It is amazing what you can accomplish if you do not care who gets the credit." This is a theme that must be ingrained into your team before watchteam backup can be effectively implemented. If someone hears something that doesn't sound right, we want—no, we NEED—that person to speak up. Perhaps 99 times out of 100, the input will be marginally useful, if at all, but when we respond to his or her willingness to participate with gratitude, the payoff on that 1 out of 100 instance will be well worth any inefficiencies, inconveniences, or damaged egos resulting from the other 99 instances. If it was easy, all organizations would have effective and energetic watchteam backup, but because it's not easy, very few do.

WATCHTEAM BACKUP ON A NUCLEAR SUBMARINE

Watchteam backup can be accurately translated to civilian vernacular as "employee engagement." Creating a team-orientated environment is no small accomplishment. In fact, you could argue that successful watchteam backup is the panacea of the effective implementation of the principles we have already discussed. Without knowledge, watchteam backup would be ineffective noise. Without standards, watchteam backup would be inconsistent and frustrating because everyone needs to be on the same sheet of music in regards to the organization's expectations. Without a questioning attitude, effective watchteam backup could not occur because the individuals' brains would not be engaged enough to recognize the necessity for backup. And, when we discuss integrity in the next chapter, you will see that without integrity watchteam backup would likely develop into office politics run amok with people vying to make themselves appear smarter under the guise of "backing up" their peers. Effective employee engagement is the result of the other principles acting in harmony.

On a Nuclear Submarine, "watch team backup" is a term that is well understood and encouraged by everyone aboard the ship. Everyone means everyone. The most junior person is expected to back up the Commanding Officer or the Officer of the Deck when his engaged brain alerts him to danger. Truth be told, quite often the efforts of junior personnel to backup senior personnel are incorrect or needless, but they are always encouraged, acknowledged, and thanked because there will come a day when that sailor's backup could save the lives of the crew. There is a trap in training junior personnel to believe that showing respect to senior personnel means deferring sheepishly to their orders. Although, we absolutely expect orders to be carried out swiftly, it is undesirable for anyone to do so with a disengaged mind. All personnel owe the rest of the crew their best efforts, and this includes engaging their minds in all of their duties, learning when vocal watchteam backup is appropriate, and preparing themselves to speak up when necessary. Speaking up when necessary, regardless of whom is on the receiving end, is called "forceful watchteam backup." As an example, assume the Officer of the Deck has just ordered the ship to maneuver right 90 degrees from due North to due East. There is a big difference between the following two reports from the Sonar Supervisor:

1. "Officer of the Deck, Screw blade noises bearing 090."

2. "Officer of the Deck. Screw blade noises bearing 090. Recommend coming further right to 105."

The first report provides the Officer of the Deck some data but does not forcefully connect the Officer of the Deck's order to drive on course 090 and the existence of a sonar contact on that bearing. The second example ties these two pieces of information in the form of a recommendation that is direct and unmistakable. This is "forceful watchteam backup."

Blind allegiance to orders is lazy, disrespectful, and dangerous. There is a term for this blind obedience: "malicious compliance." Malicious compliance is when junior personnel shut off their brains and simply do what they are told to do. This is dangerous in any environment and is often created by overbearing leaders who stomp on junior personnel who attempt to provide their input or feedback.

Consider a Nuclear Submarine at sea. Once a year, each submarine undergoes an intense tactical readiness inspection. The inspection, the Tactical Readiness Examination (abbreviated TRE and pronounced "tree") is intense and a drain on the entire crew. On one particular inspection, the results were less than what the Commanding Officer believed his crew deserved. As a result, there was a subtle, but noticeable, amount of friction between the Commanding Officer and the senior TRE member (post-command officer) on board. Once the inspection was completed, the plan was to transfer the TRE team off of the ship.

Personnel transfers are dangerous in any weather conditions. The submarine surfaces and then drives into protected waters. A small boat pulls alongside the submarine; the vessels match speed and then a rudimentary brow is used to connect the ships to allow personnel to transfer from the submarine to the small boat. The risk of someone, either the ship's crew, who are topside to assist in the transfer, or the transferring people, going overboard into the ocean is real. Therefore, prior to conducting a personnel transfer the members involved assemble in crew's mess (the space where the crew meets and also the only space large enough to brief an evolution that involves more than 15 people) for a pre-evolution brief.

The pre-evolution brief focuses on the logistics of the transfer, but it also includes a discussion of risk. During that discussion on this particular transfer, the Navigation Officer mentioned that the weather forecast predicted rough seas. The

Commanding Officer reiterated the importance of safety and deliberate action.

Fast forward to the transfer. The Navigation Officer was right. The weather was brutal. Visibility was poor and the waves were breaking over the sail of the submarine. The ship's personnel began to emerge from the forward escape trunk hatch. They moved slowly and deliberately with all of the safety gear, including lanyards that attached them to the ship. The Chief of the Boat was getting hit by strong waves. At one point, he was forced to drop to his hands and knees to stabilize himself. The Officer of the Deck was attempting to maneuver the ship to reduce the personnel's exposure to the waves, but the seas were confused and appeared to be coming from every direction. The Officer of the Deck was mentally practicing his man-overboard actions because this transfer looked like it contained a larger-than-usual risk of a man falling off the submarine.

The Commanding Officer came to the bridge. He was surprised to see how vicious the weather was. He provided the Officer of the Deck guidance on how best to position the ship to shield the personnel topside from the waves. The waves intensified. There was now water coming into the boat through the open hatches. The Commanding Officer ordered the personnel that were transferring to come topside. He was thinking that he needed to speed up the process before the weather got even worse. The small boat that came out to meet the submarine and pick up the inspection team was getting hammered by the waves and 20 foot swells.

The Phone Talker who was stationed in the bridge to relay reports and orders between the Officer of the Deck and the Control Room Supervisor spoke up. "Captain. Why are we doing this today? Someone is going to end up in the ocean in this weather."

The Captain paused. He looked around. The rain was torrential. The visibility was near zero. The top side personnel who were on top of the submarine deck attempting to prepare for the transfer were being hit with waves. The small boat that was supposed to take the inspection team was bobbing like a cork.

The Captain looked at the Phone Talker and said, "Phone Talker to Control. Order all topside personnel to get below." He then turned to the Officer of the Deck and said, "The personnel transfer is cancelled. Derig the bridge [send all of the equipment and personnel below] and rig the ship for the dive."

One hour later, the ship was submerged. The inspection team was still on board. They were angry when they heard the news that they were going to have to stay on board for another day. Many of the staff personnel ashore were perturbed at the Commanding Officer's inability to perform a personnel transfer with a "little bit of weather."

The Commanding Officer collapsed into his state room chair and shut the door. He knew he made the right decision. Those who would second-guess him, including the inspection team, didn't have the responsibility for the safety of the crew and guests. He then remembered the Phone Talker's question which he had not directly responded to. He reflected and concluded that if the Phone Talker, Petty Officer Campbell, hadn't said what he had, he probably would have proceeded with the transfer. The Phone Talker's question forced him to reconsider his environment and the risk-versus-gain of the scenario. He walked into the control room and said, "Chief of the Watch, have Petty Officer Campbell report to my stateroom."

Moments later, Petty Officer Campbell was standing at the doorway of the Commanding Officer's stateroom. Petty

Officer Campbell had been aboard this Nuclear Submarine for nearly five years. He was a respected electronics technician and leader amongst his peers, but he knew that being called to the Commanding Officer's stateroom was rarely good.

The Commanding Officer welcomed him to his stateroom and asked him to take a seat. Once Petty Officer Campbell was seated, the Commanding Officer swiveled his chair in his direction. The Captain leaned in, looked him in the eye, extended his hand, and said, "Petty Officer Campbell. Thank you." Petty Officer Campbell was slightly perplexed. It never occurred to him that his question led to cancelling the personnel transfer. The Commanding Officer held his gaze and his handshake. "In submarining, sometimes actions trump rank. Today, you were my number two and my right-hand man. Thank you. Truly."

The reality now dawned on Petty Officer Campbell. He responded with wide and proud eyes, "Yes, sir. Thank you, sir."

"No. Thank you. Today—you were the hero."

WATCHTEAM BACKUP IN THE OFFICE

When the work environment is composed of cubicles versus control rooms, the term watchteam backup is more aptly described as "employee engagement." Determining the integrity of employee engagement is very challenging and often deceptive. An exaggerated instance that we have all observed is the office "kiss up"—usually agreeing and then strategically providing "new" ideas that are often rehashed versions of the boss' ideas.

If we've all observed this disingenuous act, then it stands to reason that those in management are susceptible to seeing what they want to see and hearing what they want to hear. This is human nature. Much of leadership is identifying and then fighting our biases—some natural and some established—but all real. You take the reigns of an organization that has become complacent and suffers from disengaged workforce by encouraging the employees to share ideas and innovations—even if they have shared but been shot down in the past. It is possible, if not probable, that the first few ideas shared with the new boss are likely to come from the employee(s) considered by the rest of the group to be the "kiss up"—although the ideas themselves may have some value.

This is a difficult line to walk. The balance is so precise between providing people permission to identify every possible company misstep (knowing some people will do this only to make themselves look better) versus increasing the strength of the team by leveraging the engaged brains of the entire organization. Know now, that it is impossible to strike this balance perfectly all the time. However, the option of not trying is to live with single-point failures throughout the organization. This is not how a high performance team operates. Therefore, we must be willing to upset the apple cart from time to time in order to transition from an organization of individuals doing their jobs in a vacuum to a team-orientated environment where each person is acting as a safety net for the other.

Despite the challenges that may lie ahead of you, find solace in the reminder that everyone wants to be on a winning team. There are a few "scorched Earth" outliers in the world, but hopefully you can identify those who would put their egos before the better good. These people seek and thrive on drama, but your organization does not. Generally, drama queens (kings) are not rehab-able.

WHEN WATCHTEAM BACKUP IS ABSENT

Nuclear Submarines are an arsenal of weapons that should strike fear into the hearts of our enemies. One of these weapons is the Tomahawk missile—a long-range and highly precise weapon used to attack land-based targets. The execution of firing these extraordinarily advanced missiles is the most critical element to a successful Tomahawk strike. There are several complex variables that must be managed on a Nuclear Submarine during a Tomahawk strike mission. Therefore, submarine crews spend a lot of time training on this mission through simulated attack scenarios. These training scenarios are conducted with an intensity that can exceed an actual strike. If a Nuclear Submarine crew can excel during the simulated strikes, the actual mission typically appears simple by comparison. The reason for this is that our training scenarios include preparing for everything than can go wrong: a change in mission at the last second, material casualties that impact the ship's ability to fire the missiles, and weapon malfunctions that are possible, but highly unlikely. Create a training scenario with one or more of these contingencies and the scenario becomes intense.

The intensity of the exercise is challenging to describe. There are no discernible clues to remind any of the people in control that this is just an exercise. The ship is at sea, submerged at periscope depth with an environment designed to simulate war-like conditions to the maximum extent. The command and control unit ashore is communicating directions, orders, and changes with the same tone and tempo as it would in an actual attack. The only difference is that the weapons do not leave the ship.

To support these training scenarios, Nuclear Submarines will sometime bring a simulator, called a TOTEM (Tomahawk Test Missile), out to sea with them. The TOTEM is the same size as an actual Tomahawk missile (20+ feet long and 21 inches in diameter). TOTEM increase effectiveness of our training because it communicates directly with the submarine's Fire Control system. This allows the TOTEM to simulate malfunctions that provide the operators with identical indications that would occur in an actual casualty.

Fast forward to a strike exercise that occurred aboard a Nuclear Submarine using a TOTEM as a training device:

"In the open window, Salvo one. Fire when ready," the Commanding Officer bellowed.

"Fire when ready, aye sir," reported the Combat Systems Officer (Weps).

A salvo implies a series of launches in rapid succession. Therefore, the Combat Systems Officer began to direct a series of orders to launch missiles at exactly the correct time (within seconds). In response to these orders, the operators have to simulate most switch manipulations to prevent an actual launch of a missile. However, whenever practicable, our training program allows operators to physically manipulate switches to maximize the training value. One way to do this during Strike exercises is to physically launch "waterslugs" out of empty torpedo tubes. If a torpedo tube is empty, the tube can be flooded with seawater and equalized with sea pressure, and then the high pressure air system that would normally impulse the launch of a torpedo or a Tomahawk missile is applied to a slug of water. If the tube is empty, the operators can proceed with an actual launch with the end result of a slug of water leaving the ship at a high velocity. It sounds cool and adds to the realism of the training.

After simulating a few missile launches from Salvo One, the Weps ordered the Weapon's Control Console (WCC) operator, "In the open window, shoot Tube Two."

The WCC operator, a 10-year veteran with an impeccable professional reputation among the crew and officers responded, "Shoot Tube Two, aye" and began to simulate the launch of Tube Two.

The Weps abruptly stopped him. "Tube Two is empty" (code for we can shoot a waterslug) and continued, "Shoot Tube Two."

The WCC calmly responded, "Sir, there is a TOTEM loaded in Tube Two. Simulating the launch of Tube Two."

The Weps was an aggressive and intense leader. Although he didn't drink coffee, most would swear that he had IV stream of caffeine hooked up to his body somewhere. He also had a well-deserved reputation for being overbearing.

The Weps didn't skip a beat in response to this report, "Tube Two is empty. SHOOT Tube Two."

The WCC operator held his ground, "No sir. There is a TOTEM loaded in Tube Two."

You may able to predict where this story is headed. Instinctually, so did everyone else in the control room. The intensity and volume of the dialogue between the Weps and the WCC operator began to draw people's attention. For a Strike Exercise, the Control Room is jammed packed and includes nearly every officer. The following people were within 10 feet of the Weps: the Commanding Officer, the Executive Officer, the

Engineer Officer, the Navigation/Operations Officer, and the Communications Officer. In addition, there were at least 20 enlisted personnel in the control room. All of these other people in the control room had their own assignments that required their full attention. For example, the Engineer Officer was on the periscope as the Officer of the Deck and was positioning the ship's direction to support each successive simulated launch. However, everyone's "spidey" sense had been lit up—this was only an exercise; if there is a dispute about the status of Tube Two, why not pause the training to clarify what, if anything, was loaded into Tube Two?

However, the window for someone to provide this type of backup was of a small duration and quickly closing.

The Weps bellowed, into the ear of the WCC, "Tube Two is empty, Shoot Tube Two! That is an order."

The WCC fought valiantly to protect against the upcoming untoward event, but without the support of anyone else in the control room and his obligation to obey direct and legal orders, his options quickly dissipated. The tone of his voice was a combination of resignation and "I told you so" when he eventually relinquished, "Shoot Tube Two, Aye Sir."

Moments later the multi-million-dollar training simulator was launched into the Atlantic Ocean in search of its final resting place at the bottom of the ocean several thousand feet below.

Needless to say, the aftermath of this error was exhaustively critiqued and evaluated. How could this have happened? There were reasons that contributed to this mistake that are not relevant for our purposes, but regardless of what mistakes were made, it was clear to many people that the risk of allowing the situation at the WCC to escalate was not one worth accepting. It was only an exercise.

Any person in the control room—anyone—could have simply stated, "Stop! Training timeout. I recommend we verify the status of Tube Two prior to proceeding." If anyone had done this, the humiliating, expensive, and potentially dangerous mistake would certainly have been averted. So why didn't anyone do this? Only each person in the control room that day can answer that question for certain. However, it is reasonable to speculate that there were four main issues that contributed to the lack of watchteam backup.

1. The Weapons Officer was so assertive and confident that it was difficult to imagine that he was wrong. Never confuse confidence for competence—easier said than done. However, the skill of interpreting data stripped of the emotion with which that information is delivered is an important one.

2. The crew placed undue importance of executing the practice mission flawlessly. Although, we all recognize the virtue of sports' adage of "practice like you play," the importance of moving forward with the exercise was self-imposed.

3. Without adequate practice and preparation, the window of time available for someone to exert the appropriate watchteam backup was too small. Everyone in the control room had a sense of "*I should do something*" but the amount of time to process that information without having prepared for it adequately was insufficient. From start to finish this scenario was 20 seconds—if watchteam backup allowed for a good night's

rest to think about the matter, we wouldn't even be discussing it. The ability to recognize when the time is NOW is not innate.

4. Most significantly, the culture aboard this particular submarine clearly did not emphasize or even encourage watchteam backup. The submarine was led by leaders that would conventionally be described as "strong" in that they were direct, confident, and teeming with bravado. Make no mistake, they were good submariners with great careers and confident for good reason. However, this mishap is reflective of their failure to create a culture that fostered and cultivated forceful watchteam backup.

THE UGLY SIDE OF WATCHTEAM BACKUP

In developing a culture of strong watchteam backup, we are essentially encouraging people to vocalize missteps or potential missteps that they observe. This is tricky. Human beings are naturally gifted critics, but there is little value in criticism for criticism's sake, as Theodore Roosevelt pointed out in his famous, "Man in the Arena" speech. There is, without question, the potential for an emphasis on watchteam backup to degrade into a contest of egos and heightened senses of office politics, suspicion, and distrust. To avoid this, the concept of team must be developed first. If you find yourself in an organization that is composed mostly of individuals looking to "get theirs" instead of a group of people committed to a common goal, turning this around is a prerequisite for establishing expectations of watchteam backup. Effective watchteam backup presupposes that there is a strong team unit in place.

Creating a strong team culture and then watchteam backup among employees takes time and patience. The only exception to this process may be in industrial environments where the risk of injury or equipment damage is high. In these cases, we can, and must, emphatically and immediately insist on not only the expectation but the obligation for each member of the organization to remain engaged and be prepared to intervene, regardless of seniority, if she believes that a dangerous situation is developing. The challenge that must be overcome in these environments is one that involves the frontline supervisors. The organization leaders must, through hell or high waters, make it clear to supervisors that attempts to provide watchteam backup in regards to safe operation of industrial equipment must always be met with gratitude and never retribution, even if the backup is misguided or incorrect.

Whether implementing a watchteam culture over time or immediately, every organization will have to deal with members who love the spotlight—the "know it alls." Once you begin to discuss the importance of watchteam backup and emphasize its application, you will have certain members of the organization who use this as an excuse to flaunt their personality weaknesses. This has the potential to drive you nuts because there is no easy solution. You will want to steer clear of shutting these people down, even when you know they are only providing watchteam backup to hear themselves talk or to posture for an increased amount of attention. However, if you continue to focus on the team nature of your organization and on respectful and appropriate communication when providing backup, in time, the process will be self-correcting, but you are going to have to deal with growing pains initially.

HOW TO DEVELOP A WATCHTEAM BACKUP

The process of developing watchteam backup will differ from one organization to the next. The time and effort required to engage employees, create a balance of support versus criticism, and establish a cultural expectation will not happen on a predetermined schedule. However, the following five components of development are necessary to the process: prepare, accept questions, expect accountability, cross train and build trust.

Prepare

Those who fail to prepare should prepare to fail. Often the best opportunities for watchteam backup occur spontaneously with a short window for action. It would be nice if the opportunities for watchteam backup allowed for everyone to sleep on it, but this is rarely the case. Because of this, it is critical to have discussed this topic, over and again, with those from whom you expect watchteam backup. In fact, discussion may not be enough. There may be times when you'll benefit from role playing the process of stopping something with the appropriate level of communication. For example, you might ask a group of employees to imagine a scenario in which one salesperson overhears another salesperson misspeak to a potential customer on a routine sales call.

"All of our HVAC products will be 20 percent off between now and June 1st."

Actually, all of the company's HVAC products are **25** percent off between now and **July** 1st. What is the appropriate response from the passerby? It may be hard to believe, but practicing how to communicate in this type of situation is imperative. Without this type of practice, it is possible that watchteam backup can create an environment in which people are constantly correcting one another, and worse—in the wrong way at the wrong time.

Practicing is desirable in most institutions, but essential in an industrial environment where certain mistakes could result in damaged equipment or worse—damaged people. These are the scenarios that must be emphasized in advance. They must be practiced in advance. Not unlike the immediate actions we described in Chapter 3, when the time for immediate action presents itself, the time for thinking is nearly non-existent and our training takes over.

Accept Questions

The manner in which leaders allow questions to be asked is an important element of watchteam backup. Inadvertently, leaders can foster an environment that allows the members of the organization to turn off their "watchteam backup" switch simply through the manner in which they tolerate questions.

Assume, you are the Regional Sales Manager of a company that manufactures, sells, and distributes office paper. One of your sales associates, Dwight, is negotiating a renewal contract with your third largest client. The client has been a loyal account for years, but is now threatening to leave unless you shave five percent off the cost of paper supplies or shift his payment terms from net 30 to net 60. Consider the two scenarios:

Scenario One:

Dwight: "Jim from Blunder Stifflen Corp is not going to renew his contract with us unless he gets a five percent discount or net 60 payment terms on the new contract. What do you want me to do?"

You: "Go ahead and give him the net 60 payment terms."

Scenario Two:

Dwight: "Jim is not going to renew his contract with us unless he gets a five percent discount or a shift from net 30 to net 60 payment terms. I intend to give him the net 60 payment terms, but also negotiate a minimum amount of paper that he orders each year to counterbalance that discount."

You: "Good idea, Dwight. I agree."

Although Scenario Two is not technically a question, it is functionally a question. Both Dwight and you know that you authorize contract modifications. But that doesn't mean you have to come up with all of the answers. If you disagree with his plan, you can do so after he has presented his recommended solution. If you agree with his plan, he leaves your office with marching orders to execute HIS plan. In this case, Dwight was concerned that you might not approve either so he was able to devise a way to counterbalance the negative impact by stating his intention to negotiate an annual minimum.

This style of decision making is called "Command by Negation." People state their intentions to do something, and the leader retains the right to stop or redirect someone's efforts.

How does this relate to watchteam backup? There are two significant benefits to establishing a standard of stating intentions rather than asking for your decision:

1. It allows the decision maker to play the backup role. Presumably, the person proposing the solution has thought through his intention using the information available to him. So as you consider the merits of his intention, you are playing a backup role to a decision. Compare this to the "What do you want to do?" mentality. Although some leaders enjoy this framework

because it plays to their egos—*I make the decisions around here*—the reality is that the decision maker has now become a single-point failure. Who's backing up the decision maker in this framework? No one. Recall that one of the reasons watchteam backup is so important is to eliminate or reduce a single-point failure error.

2. By forcing people to state their intentions when faced with a dilemma or decision, you are increasing their decision-making skills. They are forced to practice weighing the pros and cons, conducting a cost-benefit or risk analysis. As a result, you are developing stronger decision makers.

Expect Accountability

Hold people accountable. When something goes south—as things invariably do from time to time—investigate what happened and why. During that exploration, identify who had or should have had information that could have prevented this untoward event. Don't limit this to the most obvious groups of people. If the event was a pump that was damaged because it was incorrectly wired, don't be afraid to speak with people other than the electricians. What did the mechanics know? The operators? The laborers? The management? You will be amazed how often we find out that someone was holding onto a piece of knowledge that would have saved the day but she didn't share it with anyone. Often times, the reason she didn't share it was because the information was outside of her perceived "channel." Watchteam backup requires busting up these lanes and reminding the organization's members that they are all in this, achieving operational excellence, together.

Cross Train

Why don't we train our marketing department on power plant operations? Why not provide our electricians with basic financial analysis training? Generally, for two reasons:

1. Management considers that idea unnecessary or absurd. Why in the world would my electricians need training on financial analysis?

2. Time and money are finite and valuable. Some managers might see a benefit in cross-training but cannot justify spending the time and money required to do so.

Cross training is always time well spent. Unfortunately, measuring its return on investment (ROI) is nearly impossible. Despite what many organizations have convinced themselves of, not every investment has a measurable and immediate financial return. However, if you are trying to lead your organization towards higher levels of watchteam backup, cross training is wise.

There are three significant benefits to cross training as it relates to watchteam backup:

1. Every second spent training someone on an area outside of her official lane of duties and responsibilities is a second reminding her that she is part of a team that expects her mind to be engaged on areas that extend beyond a narrow silo.

2. Actions speak louder than words. It is one thing to express that "we are all a team moving towards the common goal of operational excellence" and another thing to demonstrate the organization's commitment

to developing members by broadening their horizons.

3. Cross training creates improved communication between members of different functional areas. In power
production and manufacturing, there is almost always
a divide between the operations department and maintenance and repair department. The latter typically
work normal hours and the former on shift-work. The
operations department often feels isolated from the
company. Therefore, cross-training in this example allows each group to walk a mile in the others' shoes,
while also improving, significantly, the opportunity for
watchteam backup in the plant.

Build Trust

Perhaps the most challenging aspect of developing an organization with an effective cultural principle of watchteam backup
is trust. Those receiving backup must trust that that those delivering the backup are doing so from a place of genuine support for the organization's goals and performance. Further, the
individuals providing watchteam backup must trust that their
input will be taken seriously regardless of their assigned functional unit. Also, those providing watchteam backup must trust
that they will not be victims of retribution for speaking up and
potentially criticizing. (We addressed earlier the importance of
training people on how to communicate watchteam backup in
a productive, non-judgmental, and supportive manner.)

How do you build trust? There is a crucial element to discuss
before we answer this question. Trust must flow in all directions of an organizational chart to be effective. The trust that
we are discussing is two-fold and both are necessary. Trust in
your leadership and trust in the organization. Trust is generally
composed of two elements: commitment to knowledge/ability

and integrity. This composition can be loosely translated as people believe that the organization knows how to do the right things and have the moral character to do them "right." People unwittingly grade both their leader and the organization on these components—ability and integrity.

Organizational trust is best established incrementally. Erratic organizational behavior—*This month we will be making the following 28 changes*—is rarely effective even when it comes from a source of good intent. Organizational moves should be guided by the principle that is akin to the Hippocratic Oath of Medicine — "first do no harm." Organizational changes should be kept to the minimum required, but no less. Organizational changes make employees skittish and suspicious. Do not interpret this guidance as "do not make organizational changes"; the guidance is that organizational change should be kept to a minimum. It is better to make two impactful changes at the beginning of the year, than it is to make continuous tweaks throughout the year. The latter leaves people uneasy and will also result in the growth of doubt about the ability of the organization.

Trust in leadership is a topic large enough for another book, but here are a few thoughts. There are an unlimited number of ways that leaders can develop or destroy the trust people have in them. Here are two that are most directly tied to watchteam backup:

1. "People will not care how much you know, until they know how much you care." Leaders must earn trust before they can expect people to follow them. You may be a fountain of knowledge, but people would rather go thirsty than learn from someone they don't trust. Every scenario is different, but here are some ideas to consider

for establishing trust:
a. Speak directly and plainly.
b. Set attainable and measurable goals with a deadline. Reward attainment.
c. Coach before you manage and manage before you direct.
d. Always do the right thing.
e. Display your willingness to sacrifice.
f. Use "we" and "us," not "I" and "you" in your communications.

2. Develop chain-of-command loyalty. Loyalty applies both up and down the chain of command. This is a principle that is rarely discussed and, therefore, often lacking. Consider the following example of chain-of-command loyalty as it applies to watchteam backup:

A Nuclear Submarine was in the process of surfacing. Many people assume that surfacing a submarine should be simple, just drive the ship to the surface. But that is not how it works. Submarines can operate submerged because of the negative buoyancy provided by filling several large tanks (main ballast tanks) with water. This design allows the submarine to immediately become positively buoyant in the case of a casualty by using high pressure air and blowing the water out of these ballast tanks. When the emergency system is used, it is called an "emergency blow." However, when the ship surfaces, the water is expelled from the ballast tanks in a more controlled manner. During this controlled process, it is possible to have an unequal expulsion rate of water from the forward and aft tanks. This can happen for a variety of reasons.

On this particular morning, the helmsman accidentally placed a negative angle (submarine slanted with forward end deeper than the aft end). This resulted in water leaving the aft main ballast tanks more rapidly than the forward tanks (less pressure on the aft tanks). This resulted in the aft end of the ship becoming more buoyant than the forward end of the ship, which resulted in a larger down angle. Before the Officer of the Deck had time to think, the submarine had a steep down angle and was going deeper and deeper.

Holy s**t! This is a surfacing evolution, and we now 200 feet deep with a 20 degree down angle. What is going on? What should I do?

The ship control team had no ability to right the ship. The bow and stern planes (control surfaces that direct, under usual circumstances, the submarine to go up or down) were both positioned on full rise, but the ship's angle was getting steeper and steeper and the ship's depth was going deeper and deeper. The planes had virtually no ability to reverse the angle or depth because at slow speeds the planes are less effective and because the forward main ballast tanks had A LOT more water in them than the aft tanks did at this point.

The OOD ordered, "Chief of the Watch, perform a three second emergency blow of the forward main ballast tanks."

As the Chief of the Watch was repeating the order, the Junior Fire Control Man in the control room said, "Officer of the Deck, we're going too slow for the planes to bring us up. Recommend AHEAD STANDARD instead of emergency blow."

The OOD thought about this. "Chief of the Watch, STOP. Do NOT blow the forward group. Helm. Ahead Standard."

Fast forwarding through the remaining details, the ship was able to right itself and eventually surface. However, the order to increase speed by ordering Ahead Standard made the situation worse not better. It was a bad decision. A three second blow would have corrected the situation more quickly and resulted in a shallower depth excursion.

During the critique (we will discuss critiques in the next chapter), all of the members of the control room watch team were in attendance. The events of what happened and why were discussed in excruciating detail. Intending to surface the ship, but instead being hundreds of feet deep and descending rapidly is a what you might call a "big deal," so a thorough investigation was called for.

When the topic of why the Officer of the Deck belayed his order to blow the forward group and ordered Ahead Standard, he never mentioned the Fire Control man's recommendation. He explained his rationale (control surfaces being more effective at higher speeds) and assumed full responsibility for the decision. Had he done otherwise, the next time the ship was in duress and the Fire Control man had an idea, do you think he would be more or less likely to share his idea?

This is loyalty down the chain of command.

Loyalty up the chain of command is similar. Let's assume the Executive Officer informs the Commanding Officer that he intends to give the crew the weekend off and delay the maintenance and training schedule by two days. The Commanding Officer disagrees. They go back and forth, but the Commanding Officer is not dissuaded. Therefore, the crew will be working this weekend. The Executive Officer is furious.

He addresses the crew, "I know we talked about a weekend off, but the Commanding Officer insists that we stay on or ahead of our maintenance schedule; therefore, we will all be working this weekend."

This is NOT loyalty up the chain of command. This example exaggerates for effect, but how many times have you heard your boss use his own boss as the excuse for why we are doing something. Even if it is true, a leader owes his leader the loyalty and trust to disseminate the information as if it were her own.

Watchteam backup is an elusive yet lucrative dividend that is available to be harnessed in any team setting. Effectively developing watchteam backup is the golden ring that is obtained when questioning attitude, knowledge, and standards are successfully combined. In theory, it eliminates single-point failures in our organization. In practice, it reduces these single-point failures substantially by joining all of an organization's members to its operational success. Further, it empowers all members of the organization to speak up, and it teaches them when and how to do so, to ensure that the full capacity of an organization's resources is leveraged.

CHAPTER 6

INTEGRITY

The roof of a building is the component that is seen least, but whose breach would be recognized first. The roof protects inhabitants from the elements such as inclement weather and blood-thirsty carnivorous flocks of birds. These events are unpredictable but when our building's roof is strong, we proceed with our work as usual. Without a roof, you don't even have a building (we're sure a working definition of a building would include a reference to "enclosed") and with a weak roof you are vulnerable to nature's unpredictable and unforgiving behavior. Integrity is your organization's roof.

Integrity is vital to a Nuclear Submarine's performance and survival. The strength and resiliency of the principles that we have discussed all depend upon integrity. So why then are we describing it last? The other principles discussed in the previous chapters can be woven into the fabric of your organization with results that are, to an extent, measurable or observable. These principles increase performance, productivity, and morale. Despite your best efforts, the successful implementation of these principles relies on external variables that are so fundamental, we take them for granted. Examples include air to breath, water to drink, and food to eat. Integrity falls into this category of variables with one notable anomaly: it is an internal, not an external, factor. You can, and must, demand the highest levels of integrity in your organization if you have any hope of surviving the unpredictable yet inevitable storms that your organization will endure.

Air, water, food . . . and integrity are often taken for granted because we make the (faulty) assumption that they exist without effort. Recall, the roof of the building is the one that is seen least.

WHAT IS INTEGRITY?

Integrity and honesty are two words that are related but are incorrectly used interchangeably. When junior sailors hear about the standard of integrity in our submarine community, they often think about honesty—simply telling the truth. They may think, "Okay, I get it. I'll be honest. That's easy." How cute. Maintaining your integrity is extraordinarily challenging for anyone and, although counterintuitive, becomes more difficult the more seniority you have.

There are elements of honesty baked into integrity, and having an "honest" organization is important, but there's more.

So what are we driving at?

The Nuclear Submarine community uses an unofficial definition of integrity, whose origins have been attributed to multiple people including C. S. Lewis, that captures the essence of the organizational standard of integrity, while also highlighting the subtle difference between integrity and honesty.

> *"Integrity is doing the right thing,*
> *even when no one is watching you."*

Integrity is essential in highly technical and industrial environments because the equipment and systems used are dangerous and failure can be fatal. A submerged Nuclear Submarine operating in the world's harsh and unforgiving ocean environments certainly falls into this category. Aside from the more obvious enemies that a Nuclear Submarine goes head to head with, the ocean is probably our most formidable one. The ocean is always trying to penetrate the hull of the ship through its weakest point. If there is breach of watertight integrity, the ocean will find it. It is a formidable enemy vigilantly waiting for a momentary weakness.

To put this into perspective, there are over 30 hull penetrations that are routinely opened in a controlled process, such as taking in seawater to convert it to drinking water or using seawater to cool engine room components. Each of these penetrations represent a palpable opportunity for an operational mistake to result in flooding and the loss of the ship and crew at sea. This risk looms large on every submariner's mind. As a result, every submariner relies on everyone one of his shipmates to do the technically correct thing—even when no one is watching. This standard is absolute and there is little, if any, relative component applied. There can be no small breaches of integrity on a Nuclear Submarine.

Given the highly technical nature of a Nuclear Submarine's operations, this standard of integrity is most often applied to how to make decisions when faced with a technical issue—large or small. For example, consider a mid-grade, nuclear-trained mechanic performing maintenance on a primary valve ("primary" indicates a system that houses radioactive or potentially radioactive fluid components) at 0200. The mechanic observes three drops of water leak from the valve and onto the floor. There are a few things he could do at this point.

It may not be simple to identify what is *ethically* correct, but the technician knows what is *technically* correct. He could just wipe up the drops, throw away the rag, and move on with life. However, that action is not technically sound, correct, or consistent with his training. Through his training, he is expected to announce loudly "SPILL, SPILL, SPILL" and then report this to the controlling station in the engine room. "SPILL" is a code word that will set in motion a protocol of pre-planned actions from a host people (recall it's 2 a.m.) to ensure that the situation is addressed with the most robust technical standards. The actions will ensure that the mechanic has no radioactive contamination on him or his clothes, and if he does, actions are taken to remove this contamination and prevent the spread of that contamination. The actions also include ensuring that the valve being operated is restored to its full strength condition, the area is cleaned up in a manner that ensures there is no risk of the spread of contamination, and the materials used for the cleanup are controlled as potentially contaminated and securely separated until they can be disposed of properly at a shore-based facility. These actions will require, at a minimum, the effort and oversight of at least 10 people most of whom were probably enjoying the most precious and rare commodity on a Nuclear Submarine—sleep.

In this scenario, there is a 99.9 percent probability that there would be no adverse action to any person, the nuclear power plant, the submarine, or the environment if this mechanic did not yell "SPILL." However, our technical procedures are written to create and protect an extremely large safety buffer to prevent any damage or danger to the submarine, its crew, the environment, and the public. The mechanics knows through his training, beyond a shadow of a doubt, that he is expected to take the technically correct action. Period. The Nuclear Submarine's safe operation relies on upholding this standard of integrity.

The Navy accepts people from all walks of life from every corner of the country. The consequences of the variations of backgrounds and upbringings extends beyond whether one refers to soda as "pop" or "cola." There are fundamental character differences that are deeply rooted in a person's upbringing such as the definitions of "right" and "wrong." Regardless if the sailor was from the roughest neighborhood in Detroit, the poshest neighborhood of Beverly Hills, or a farm in Kansas, if a submariner says that "x happened," when describing a technical or operational event, we all believe that "x happened"—we must. However, we must also maintain our questioning attitude. This is where the interaction of the five pillars is critical. Further, a submariner can be trusted to perform his technical duties with the same rigor and procedural compliance unattended or supervised by the Commanding Officer. Submariners need to know this about one another because without this trust, no one could sleep and the safety of the ship would be at risk. If we observe any indications that this is not true, that person in doubt is likely to be left at the pier at the next port of call or launched out of a torpedo tube at sea. (Although sometimes tempting, we rarely actually launch people out of torpedo tubes.) Each member of the crew has the opportunity to earn this trust through a rigorous qualification process that starts when he walks aboard the ship. It is during this process where peers and supervisors observe and test the knowledge, skills, and integrity of each crew member before he is entrusted with the responsibility of performing his duties without constant supervision.

WHY SHOULD I CARE?

Is there anyone that would say, "I don't care about the integrity of my organization?" Probably not, but there is probably someone that would say, "I don't care about my own integrity." These people would be difficult to find, but only because of their willingness to state that which is socially frowned upon. Most of us were taught at a young age that lying, stealing, and cheating are "wrong" but that doesn't mean that we abide by these rules at all times. Fortunately, we're equipped with a Freudian superego to provide us a healthy dose of guilt when we don't follow the rules and dissuade us from making these infractions a habit. Of course, there are some without a superego, or a broken one, and they lie, cheat, and steal without remorse.

Our first point is that you care because you know, intuitively, that you "should" care. Unfortunately, when it comes to expending energy and resources to develop a high-performance culture for your organization, "should" is not enough. In fact, "should" may be the most dangerous word in the English language because it implies expectation without action. A famous psychiatrist Albert Ellis developed a form of psychotherapy called rational emotive therapy. He would warm his patients against "shoulding" on themselves.

"Should" is not enough.

The reality is that your industry will largely determine the extent to which you expend effort shoring up the roof of your building—your organization's integrity. For example, let's say you run a restaurant and the procedure for cooking French fries is that the potatoes should be in the fryer for between 4 – 5 minutes. It's a Friday night, the restaurant is packed, and the kitchen is buzzing with controlled chaos. The cook pulls the fries out of the fryer after 3 minutes. It wasn't just an honest mistake. She knew the correct procedure but she intentionally deviated from the technically correct process.

If you are the general manager of the restaurant, do you have a problem? Maybe. Perhaps it depends on how often cooks are comfortable deviating from procedures and recipes. But, if this instance came to your attention would you shut down the restaurant, ask the patrons to leave, fire the cook, and then hold a two-day training session on integrity. Probably not.

What if you are operations manager at a toy manufacturing plant. In the last 30 minutes of the backshift, an operator notices that one machine is operating outside of its allowable tolerance band. As result, one of the toy components is being produced in a manner that will not be compatible with the remainder of the toy. As an example, the doll's arm is being produced disproportionately small. The operator knows that when this happens, an alarm is supposed to go off and then automatically shut down the line. Further, he knows that it is his responsibility to manually shut down the line if the alarm malfunctions. He chooses not to because he does not want to delay getting off work and a line shut down could hold him over past his scheduled shift time. "Let the next shift worry about it." The operator's lack of integrity resulted in, at best, a waste of material and time and, at worst, a dysfunctional toy in the market place. Do you have a problem? Yes, you do.

Assume that the disfigured toy is traced back to the operator that first identified the malfunction. Further assume that the operator continues to breach his integrity by claiming that he misunderstood the expectations of the operator. In response, you may choose to shutdown the plant for a day to conduct refresher training on the correct process for monitoring and taking action for out-of-tolerance concentrations. That training is expensive. The plant shutdown brings operating revenue to zero. You are probably going to have to pay your back and midshift workers overtime to attend the training. Further, management is likely to stick you with a science project of reviewing the logs for all of the production done on that person's shift, if not the entire plant's logs.

If you were doing the math in your head about the cost of those corrective actions, there is a relatively large dollar figure swirling around in your mind. Many of the "in-house" corrective actions often bark up the wrong tree. That operator and all of the others know what the technically correct answer is, but this one individual chose not to do the right thing. Conducting operational training is never a "bad" thing. However, if the mishap was caused by an integrity violation, the operational training may be a costly corrective action that isn't addressing the heart of the matter.

These two examples serve as the ends of the spectrum of the "why" integrity is important. On the one side, a restaurant may deliver inconsistently prepared meals, which could reduce patronage over time, on the hand people could die (or the organization spends millions of dollars because of an integrity violation).

Integrity Bleeds into Principles

Many of the other "whys" for integrity bleed into the principles we have already covered:

1. **Learning organization.** You've spent a good chunk of change to develop a training program for each of your company's 73 sites across the nation. You flew all 73 general managers to your headquarters in Chicago in an effort to "train the trainers." You emphasize that the tests that follow each of the modules must be proctored and closed-book. This is essential because you need real test results to measure the effectiveness of the training. You trust the people in the room to run their sites somewhat autonomously. You should be able to trust them with specific direction about test taking.

Thirty percent of the GMs do not enforce proctored or closed book exams. The collective data your team evaluates is garbage. You make decisions based on the results of these examination scores. The evaluation was a waste of time and money and your decision has the strategy of throwing a dart—all because nearly one-third of your GMs lacked the integrity you assumed they had.

2. **Procedural compliance (standards).** You've spent the last three years standardizing your manufacturing and distribution processes by creating and implementing procedures. You've spent just as much time training supervisors that procedural compliance is now a "standard" and that your expectation is that each supervisor communicates this clearly and repeatedly to employees. You direct each supervisor to observe an employee performing a task using a procedure by following the audit guide your team created. The Instrumentation and Controls (I&C) Supervisor chooses to "gun deck" (fabricate) these audits. As a result, the I&C department's procedural compliance fails to become a standard, and they continue to make the same costly errors they had in the past.

3. **Questioning attitude.** At a coal-fired power plant, the site administrator performs a daily set of functions that allows him to order the correct amount of coal for the following day. This process involves reviewing the amount of coal burned the previous day, measuring the amount of coal that is in the plant (loaded in the bunkers), walking the coal yard to assess the amount of coal outside, and reviewing the weather report tomorrow. At the end of the day, he orders the number of tons for

the next day, which goes through the company's ordering system.

The procurement officer who handles all procurement, except coal, notices that the order for tomorrow is only 10 percent of what she would expect it to be—looks like a typo, one less zero than John intended. The company guidelines clearly state that the procurement officer reviews and approves all of the procurement orders each day. However, coal is an anomaly and her approval is administrative only. No one is going to hold her responsible since John has full coal yard responsibility.

She has asked John to teach her the coal ordering process, but he has refused—time and again. Jane knows that if she doesn't do anything there is going to be a maelstrom at the plant tomorrow. Her computer's cursor slowly pulses over the "approve" button, her finger hovers over the left click button as she eyes the speed dial button to John's line on her phone.

With one call, she can fix this. "Screw it," she thinks. "It'll be John's ass on the line not mine and besides he deserves it for refusing to involve me in the process."

Click. Approve.

The next day, the plant loses tens of thousands of dollars because there is not enough coal to burn, and they have to switch the boilers over to the more expensive natural gas.

4. **Watch team backup (employee engagement).** You are the VP of a large Eastern Region real estate

development firm. In the Baltimore office, the assistant project manager (APM) notices that the preferred mechanical contractor, AA Best Contractors, performs subpar work with ineffective and apathetic supervisors. She has addressed this problem with the project manager (PM) several times but the PM is dismissive of her comments and insists that the contractor is doing just fine.

On a day that the PM is on vacation, the owner of a different mechanical contractor, ZZ Best Contractors, calls. He wants to speak with someone about the competitiveness of his most recent bids and what he can do going forward to be more competitive. The APM informs him that the PM handles all of the bids, and he should call back next week.

Curiosity, gets the best of her. She pulls the PM's binder on their most recent project and finds the "Bids" tab. Her jaw hits the ground when she sees that ZZ Best bid 10 percent less than AA Best, offered stronger performance guarantees, and had substantially more credible references than AA Best. She places the binder on the desk with care as if she was handling a crime scene weapon. She collapses into her chair, her arms drape over the sides, and she slides into the leather executive chair so that her back is nearly resting on the seat. She stares at the ceiling and exhales loudly. Bits of data that had previously resided on the outer edge of her radar screen start to coalesce into a clearer but dirty picture—the new Mercedes, the silk suits, the glitzy watches, and his current vacation to the Caribbean. Something wasn't right.

Unfortunately, she never mentions any of this to you. You and the PM appear to be pals. When you visit, there's always golf, dinner, and baseball games. If this is a "kickback" situation, she fears that maybe you are in on it too. Blowing the lid on the situation would then jeopardize her own job. She keeps her mouth shut, but also feverishly works on updating her resume.

Not only are each of these scenarios financially costly, they provide a mirror into the organizational culture that is either headed into danger or one that will have a difficult time surviving unexpected storms—the times when the only successful solution will be to pull together as a team in the same direction.

CAN INTEGRITY BE TAUGHT?

It may be reasonable to suggest that people can be taught integrity. Definitions of "right" and "wrong" can be calibrated. Maybe that is true. However, when it comes to the operation of a Nuclear Submarine, the Navy weeds out anyone who demonstrates a breach of integrity, of any magnitude, in the training pipeline which is nearly two years long. To describe this training as academically challenging would be an understatement. The failure rate among enlisted sailors is often 50 percent or greater. Surprisingly, the vast majority of these failures are for integrity violations—inside and outside of the classroom.

Consider the following true story:

Fifteen sailors formed a softball team to compete in the Orlando Navy Base league. These players were nearing the end of the classroom phase of their training at Navy Nuclear Power School. Describing this school as intense and challenging is an understatement. There are eight hours of classroom instruction followed by up to four hours of homework and studying every day for six months. Although this is a school for enlisted sailors, anyone who could graduate this school, could excel at the nation's most prestigious colleges. These were the Navy's best and brightest—the Navy's future nuclear power trained submariners soon to be among the "nukes," as the rest of the crew would lovingly call them.

The guys on this softball team had lived through an intense training program together—Basic Training (Boot Camp), A-School (Technical school for mechanics, electricians, or electronic technicians) and now Nuclear Power School. The team of young, bright, and motivated sailors was almost unbeatable on the softball diamond. After a summer-long season, the team breezed into and through the playoffs directly to the championship game. It was an exciting time of life for these sailors. Nuclear Power School graduation was less than two weeks away and they were on the verge of winning the Orlando Naval Base Commander's Cup. After an extra-inning game in the unmerciful Florida summer sun, they were victorious. When the winning run scored, the team jumped up and down, hugged each other, and threw their gloves in the air. An outside observer might have been fooled into thinking there was some special significance tied to the game.

After their exuberance died down and the equipment was packed up, the team chose to go out to eat and celebrate its victory at the high-brow establishment of Pizza Hut a few blocks away from base. This Pizza Hut served beer and the guys on the team that were old enough to do so ordered themselves some. (Underage drinking was considered an integrity violation, so was providing alcohol to those underage if you were of age.) The waitress delivered those of age their beer—in pitchers. Drinking or not, this was an excited group of guys enjoying the fruits of their victory—hot pepperoni pizza with a cold glass of Pepsi or beer. There is no doubt this group was acting rowdier than the average Pizza Hut family.

Their memorable day was about to become memorable for a different and painful reason.

Also at this Pizza Hut was an Ensign (the most junior Naval Officer rank), who was a new instructor at the Nuclear Power School. No one knows exactly what he saw; there are two sides to every story and the answer is usually somewhere in between. Nonetheless, he walked to the team's table, showed the group his military ID and demanded to see everyone in the group's ID. He suspected underage drinking. When he confirmed that most members of the group were under 21, he called base police.

Less than a week later, each and every player on the team attended Captain's Mast (non-judicial punishment). The rules of our judicial system do not apply. One man, in this case, the Commanding Officer of Nuclear Power School, reviews the evidence, makes a decision and administers punishment, if appropriate.

The Commanding Officer expelled every single member of the team. Based on the Ensign's observation and statement, the Captain determined that there was underage drinking at that table and that everyone at the table knew or should have known it was occurring. They were all found guilty of "integrity violations" and disqualified from submarine service. They were all sent directly to the fleet (non-nuclear powered surface ships) to serve the remainder of their contracted time.

The second baseman had been accepted for a Reserved Officer Training Corp. (ROTC) program and was scheduled to attend

Ohio State immediately after graduation. The right fielder had been accepted to the Naval Academy Preparatory School. The first baseman ranked 4th academically in the class of 200. The officer programs were taken away and the academic successes were in vain.

The Commanding Officer chose to accept the risk of false positives in order to protect the fleet. His decision may have been unfair for these eleven men (who-drank-what at that table is unknown to us), but there are thousands of submariners at sea who depended on the Captain to send them men whose integrity was squeaky clean. One chink in the armor of integrity was one too many for the Commanding Officer of the Nuclear Power School to accept. This is harsh by any standard.

Reasonable people can disagree when discussing the correct response to a breach of integrity. There are some people (many of whom are in the Nuclear Submarine community) that believe that once you've broken your integrity there is no remedy. This is not unlike a personal relationship that has been damaged because of a lie. *I'm not sure if I can ever believe you again.* Such a visceral response in a relationship is easier to understand because of the intense emotions, including the expectation of unconditional trust. The emotional bonds between the crew on a Nuclear Submarine are relatable to those of an exclusive intimate relationship. However, there are others who would contend that if a junior sailor that is caught and punished (both formally and from the informal social shunning that is likely to occur) for an integrity breach, he may quickly learn his lesson.

So while there may be variances regarding the response to a first-time integrity breach, the Nuclear Submarine community is very much united in a "two strikes and you're out" philosophy. A third chance is less common than a unicorn farm.

Back to the original question: Can integrity be taught? The answer is yes. However, the more important question is how many times does an organization allow someone to fail the final examination. Outside of sociopaths, people can be taught integrity, but some people may require more detention and summer school than others. Can your organization withstand that burden and risk?

IS PERFECTION ATTAINABLE?

No.

We are both several years over the horizon from our careers in the Nuclear Submarine community. Therefore, it is human nature to be tempted to remember the good better than the bad. Despite all of our efforts from day one of training to the day of retirement to reinforce this standard of integrity—it was never perfectly abided. There were more than occasional instances when people tried to cover up their mistakes, and in doing so, not telling the full truth about an event or mishap. No one is perfect, and therefore no organization is perfect. Even we, the authors, shaded the reality of a situation to our own benefit on a few occasions—using paper thin rationalization to do so.

It is delusional to think that "We are all Boy Scouts. We are always honest. And we always do the right thing." Unpleasant things do occur on submarines. As an extreme example, four months into a Western Pacific deployment, one control room watchstander stabbed another. During the investigation, the truth was very challenging to discover because each person's story (including the witnesses) varied greatly. These scenarios are handled with the utmost of attention and aggressive disciplinary action (although walking the plank is an option that is not a viable one aboard a Nuclear Submarine).

Given the intense rigmarole previously shared about "everyone trusting everyone," how does the Nuclear Submarine community deal with the knowledge that this mantra is not perfectly correct?

There is no perfect answer to that question, and we recognize the irony in that response.

However, there are tools that you will have to identify and then hone your ability to use them. The latter is much more important than the former. We are about to give you three tools but these tools are only your running shoes. To complete a marathon, you must practice, train, and endure hardships—that's the part that matters most. You will have to practice with the same fervor that you would practice putting, seventh chords on a guitar, or public speaking skills.

Undoubtedly, much of your practice will be performed with your actual work within your organization. We encourage you to maximize the amount of virtual practicing you can complete. This means mental preparation, execution, and evaluation—all performed away from the arena of your organization. Successful athletes are taught the art of guided imagery in which they allow their minds to think through every detail of a competition. The difference between an Olympic athlete gracing the cover of a Wheaties box and living a life of obscurity can be one-tenth of a second—competitors will gain every edge they can.

The same is true of your challenge. You don't have the luxury of a two-year training pipeline that thins the herd by 50 percent to ensure you have only the best, brightest, and most honest people. The "correct" strategy to increase your organization's integrity is fickle, and the sweet spot on a spectrum of philosophies is razor thin. The Hippocratic Oath, "first, do no harm," is worth considering because it is very easy to think that you are raising the integrity standard of your organization when, in fact, your actions are having the opposite effect. Recovering from a mistake such as this may not be impossible, but it can be in the same zip code of impossible. This is why we want you to work through these challenges and solutions in your mind before implementing them.

There are three tools, among many, that stand out as worthy of exploring in the effort to maintain the highest standards of integrity on a Nuclear Submarine.

1 . Don't Shoot the Messenger

No one likes to receive bad news. But let's get one thing straight at the onset. If you deal with bad news poorly, there's good news and bad news. The good news is that you will receive less bad news, and the bad news is that the good news will be an illusion. The bad news will still be out there, but you won't know anything about it until it finds you. Worse than that, you will inadvertently create a culture that hides information at all levels. A reasonable person may disagree about the merits of the "trickle down" economic theory, but only an unreasonable person will dispute the "trickle down" organizational culture theory.

It's not complicated. If your tear the head off of your direct reports every time they provide you with bad news, in time, they will do the same thing with their direct reports because they know what lies ahead of them in relaying the information. Often, this trickle down theory is occurring beneath a person's conscience, but at the end of the day, humans too can be trained to salivate when a bell rings. When an organization stops pushing bad news up the food chain, the tiles of the organization's roof start to fly off. Integrity is the first casualty of this unfortunate, but not uncommon, phenomena. Anything that resembles "bad news" will either push everyone to turn a blind eye to it or to grab the brooms to sweep the news under the rug. Both of these behaviors hit integrity directly in the heart.

Bad news hurts more than good news heals, so we are almost always living with an emotional "bad news deficit." As a consequence, when you have a wheelbarrow of manure deposited in your office, it is difficult to have a grateful demeanor and calming influence to the delivery man. If building a culture of integrity is important to you, you must learn to do exactly this. Yes, really! In fact, with practice you should be able to receive the news as though you were expecting it. The messenger's heart rate is much higher than yours, and your first challenge is to calm him down. The calmer he is, the better and more credible information you will receive, which ultimately is the key to preventing bad news from becoming awful news. Further, the organization will benefit from the trickle down theory of organizational culture. Because the leader knows how to handle bad news, so will the supervisors and ultimately the rest of the organization.

Consider the following reactions from two different Commanding Officers on Nuclear Submarines.

The news: The Engineering Department on Nuclear Submarines with a certain type of high pressure gauge glass (a visual water level indicator) installed in a primary system (a system that contains radioactive or potentially radioactive fluid in it) is directed to replace or reinforce the glass. The community's technical support group has identified a design flaw that results in a chance, although unlikely, that the gauge glass could break, and, because of the large amount of pressure in it, the "break" would be more of a "blowout." As an aside, this technical group is a bunch of smart people at desks evaluating submarine technical data. The results of their evaluations always seem to create more work for the Engineering Departments of Nuclear Submarines. There is no doubt that they play an integral role in the continued safe and operationally excellent Nuclear Submarine community. But how many people do you know that react with joy when a significant amount of additional work lands square in their lap because of an "engineer's evaluation"?

The environment: The Nuclear Submarine is scheduled to leave port in two months for a seven-month deployment. The months leading up to a deployment are extraordinarily stressful. The quantity of items to perform, plan, and prepare are innumerable. This does not include the unexpected crisis that always rears its gremlin head during this period. Submarine deployment dates are not scheduled as "on or about" a certain day. The submarine community expects submarines to deploy on the scheduled day, if they don't (regardless of the reason) the Commanding Officer's reputation takes a hit.

In this case, the gremlin taunting the Commanding Officer is the failed drug test of three of his most talented sonar technicians. A failed drug means immediate discharge from military service—zero tolerance. However, the ship cannot deploy down three sonar men and the Commanding Officer is working fervently to find replacements with experience and talent commiserate with those who just left his ship. This is proving to be a Herculean task as the submarine community doesn't hold a bench strength for situations such as these. The Commanding Officer is writing a passionate email to the detailer (the broker of submarine qualified personnel) explaining the necessity of these replacements, when . . .

Case 1: The Engineer Officer knocks sheepishly on the Commanding Officer's open stateroom. The Commanding Officer hears the knock but continues to violently punish his keyboard. A minute or more passes, the Engineer Officer waits patiently to be acknowledged. The Commanding Officer stops typing, exhales loudly, turns around and exclaims, "What?"

The Engineer begins to explain to the Commanding Officer the summary of the technical memo. The Commanding Officer displays no indication that he is paying much attention to the Engineer until he hears the words "required modification," at which point his eyes launch laser beams into the Engineer's.

Eng, I am sick and tired of all of these incessant and unnecessary modification orders. I am not interested in appeasing every do-nothing engineer across the country, I'm trying to get this submarine deployed on time, and I would appreciate it if you choose to help me. Your job is to make sure the engine room is in full strength condition to make electricity, propulsion, water, and air for seven months! Do you think that you can do that or are you going to keep bothering me with technical memos instead?

Yes, sir. I can do that. However, please be advised that the memo states that the upgrade should be done within three months.

Eng, did you not hear what I just said? You tell them that our schedule precludes us from making that modification until we return from deployment. Understood?

Yes, sir, says the emasculated Engineer Officer respectfully with a not-so-subtle undertone of disgust.

Fast forward four months . . .

During a routine engineering drill session (practicing responses to various casualties), gauge glass referenced in the memo fails. More accurately, it blows out. The water behind this gauge glass is pressurized to hundreds of pounds per square inch—to put that amount of pressure into perspective, a door cannot be opened by the world's strongest man with ONE pound of pressure per square inch on it. As a result, several people suffered shrapnel-like injuries due to the force of the shattered glass. Also, potentially radioactive liquid began to spread throughout the engine room. After a two-day, all-hands effort, the spill is cleaned up and, as expected, no contamination was created or spread. However, the ship was required to return to port to replace the gauge glass and provide medical care to the injured personnel.

Limping into port, midway through deployment because of a preventable material deficiency is an order of magnitude more humiliating than delaying a deployment to make a required repair.

Of course, there was no opportunity for Engineer Officer to tell the Commanding Officer, "I told you so." In fact, the opposite was true; the investigation revealed that the Engineer Officer was deficient in his duties because he did not make the modification mandated by the technical memo. The Commanding Officer claimed that although he was briefed on the modification, he did not recall the Engineer Officer informing him about the required time frame for repair.

Case 2: The Engineer Officer knocks on the Commanding Officer's open stateroom. The Commanding Officer hears the knock but continues to violently punish his keyboard. A minute or more passes, the Engineer Officers waits patiently to be acknowledged. The Commanding Officer stops typing, exhales loudly, turns around and with a big smile on his face and says, "Good morning! How can I be of service to the world's greatest Engineer Officer today?"

The Engineer Officer laughs and begins to explain the technical memo. The Commanding Officer interrupts him and asks him to sit down (in the only other available seat in the Captain's stateroom). As the Engineer explains, the Commanding Officer listens patiently, diligently, and fully. When the Engineer finishes his brief, he waits for a response from the Commanding Officer's who is now staring at the ceiling with his hands gripped behind his head.

What do you think we should do, Eng?

I think we should get started today.

"Are we going to delay our deployment date if we do?"

The Engineer Officer looks at the floor pensively for a few seconds and then replies, *Maybe. But if we start immediately, maybe not. I respect the sanctity of the underway data, but I shudder to think about this gauge failing during deployment. Then what?*

Eng, work with the Chief Mechanic, put a repair plan together and then brief me at 1600.

They choose to make the modification prior to deployment. The submarine's deployment date was delayed by one week. The Commanding Officer suffered a black eye on the waterfront with many armchair quarterbacks questioning why he didn't just wait until after deployment to make the modification.

It is very possible that both of these Commanding Officers are equally likeable people outside the hull of the submarine. They are both good husbands, fathers, friends, and citizens. The life of a Commanding Officer can be as stressful as any other assignment on the planet. The first Commanding Officer's response makes him sound like an evil man and a poor leader. If he was either, he would never be in command of a Nuclear Submarine. The second Commanding Officer's response makes him sound like a stress-free carefree leader. If he was either, he would never be in command of a Nuclear Submarine.

The difference in their responses is likely the effort they have put forth in preparing for dealing with bad news. If no work is invested in developing pre-planned responses to receiving bad news, you put yourself and the organization into a very bad position. Your response will be reactive and inconsistent because it will be reflective of your mood that day. Recall, the reason you want to be approachable is to prevent problems from being hidden from you and then in turn lowering the integrity standards across the organization. This is too important to trust yourself with. You can, however, trust yourself to execute a pre-planned response. One that you must practice in your mind, over and again, until the response becomes second nature.

There is no universe in which you will not be delivered bad news. This is going to happen and you know this. Additionally, this news will almost always come at the worst time. We are not discussing an "if" scenario; we are discussing a "when" scenario. So get yourself prepared for the "when."

As we like to say in the Nuclear Submarine world, "Don't cross the street to get your ass kicked."

2. You Get What You Inspect Not What You Expect

The father of the Nuclear Navy, Admiral Hyman G. Rickover, was quoted as saying, "You get what you inspect, not what you expect." From a leadership perspective, the contrast between those two verbs is striking. In fact, there are people in the Nuclear Submarine community that are averse to that expression because they feel that the saying implies, not so subtly, that you can't trust people to do their jobs. Believers would counter this cynicism by pointing out that the inspection process does not convey distrust, it conveys interest. On a Nuclear Submarine, as well as in many other environments, a leader's time is his professional currency. Spending that currency by reviewing the work done on a particular item, sends a very clear message: "I care about this, and therefore I expect you to also."

Leaders do not have the time to inspect every aspect of operations that they are responsible for. Therefore, although it takes only a few seconds for a leader to re-iterate to a team the expectations of procedural compliance, the process of observing an evolution for its adherence to procedural compliance may suck up several hours. As a testament to Admiral's brilliance, there is always a gap between expectation and reality that is quickly identified through inspection. Always. Further, this gap can never be bridged by talking. Employees learn quickly that they will hear all types of guidance, philosophy, and expectations from management. If they were to jump through hoops each time they heard something new, they would likely risk their jobs because there would be little time remaining for the basic core requirements of their jobs. Optimistically, the employees aren't learning to disregard guidance from management, but they do learn to push the information they hear through a filter that is typically askew from the intended message.

This discussion has a flavor of increasing operational effectiveness by bridging the gap between management standards and deck plate operations. This is true. But we are discussing tools to increase the standards of integrity in an organization. How do they relate?

On a Nuclear Submarine there are, conservatively, 500 gauges and meters that are required to be calibrated each year. Typically, the boat designates, as a collateral duty, a mid-grade enlisted sailor to be the "Gauge Calibration Petty Officer." This duty does not include physically calibrating every gauge and meter on the ship. His responsibility is to administratively track the ship's progress, evaluate the ship's process for calibration, and verify that the correct technical procedures are being used. Each week, the Petty Officer prints out a report that includes metrics about the program. Most senior leaders will skip right to the final line item: "Percent of calibrations within periodicity." Any number below 90 percent is generally reacted to negatively.

On one particular Nuclear Submarine, the weekly report was reviewed by the Engineer Officer and his reaction was adverse, to say the least, when he saw that the ship's percent of gauges and meters in periodicity was at 67 percent. Clearly, this number didn't shrink overnight but with all of the shiny objects competing for his attention, it had been several months since he had thoroughly reviewed a report. He immediately sent for the Gauge Calibration Petty Officer. They sat down in the Engineer Officer's stateroom to review the report. The Engineer Officer asked the Petty Officer how long it would take until the program could be above 90 percent. The hard-charging, ambitious, and eager-to-please Petty Officer responded that he thought that with an aggressive effort, they could be above 90 percent in two months.

"Two months?! How about two weeks?"

The Petty Officer's face displayed both surprise and confusion. "Sir, there's over 100 gauges and meters to be done . . ."

"Where there is a will, there is a way, young grasshopper. I believe in you. You have two weeks."

After an awkward pause, the Petty Officer responded, "Yes, sir. I'll get it done."

Two weeks later, the number of compliant meters rose from 67 percent to 95 percent. The Engineer was impressed. He was impressed by the Petty Officer's efforts to meet and exceed the aggressive timeline. He was more impressed with his leadership. He inspired a young sailor to accomplish a task that he originally estimated at two months in just two weeks. He could barely refrain from literally patting himself on the back.

Three years later, that same Engineer Officer (we will call him Mike) was assigned to the Nuclear Propulsion Examining Board. This Board is a team of Naval Engineering professionals whose task is to observe and audit the safe operations of Navy Nuclear ships. This group's arrival on a Nuclear Submarine was one of the most stressful three days of the submarine's calendar year. The inspection was fast-paced, comprehensive, and a strain on the crew. One of the elements of the inspection was an evolution period. During this period, each member of the inspection team was assigned to monitor and grade the performance of a routine maintenance item.

On one of his first inspections, Mike was assigned to monitor a temperature gauge calibration. Several members of the inspection team joked with him, "You can record that grade now." Mike didn't understand. The other inspectors began to share their stories about observing gauge calibrations. Each story worse than the next. One of the senior inspectors joked, "I don't think we've ever observed a successful calibration." Mike remained confused. "What's so difficult about a gauge calibration?"

After watching three mechanics attempt, unsuccessfully, to calibrate a simple temperature gauge, Mike returned to the inspection team's room to write his report. The procedural violations, the faulty test equipment, and the general lack of knowledge made the grading of the evolution easy: "Unsatisfactory." But that's not where Mike's mind was. Mike's thoughts were teleported back to three years ago. An event that he had once considered reflective of his strong leadership abilities was immediately flipped into one of his largest leadership failures. Truth be told, prior to that day, Mike had never observed a gauge calibration. He had no appreciation for its complexity or length. Even if the group today had mastered the evolution, it still would have required an hour to complete.

Recalling, the events of three years ago, Mike realized that the combination of his ignorance, the Petty Officer's eagerness to please, and Mike's "strong" leadership style mixed a concoction that directly contributed to a breach of integrity. This breach was not limited to the Petty Officer. Recall, that he was not responsible for calibrating the meters, only coordinating with the respective divisions who owned the meters. Mike was sick to his stomach that he had put so many young and impressionable sailors into such an unfair situation. What type of longterm damage did he cause to the definition of "integrity" that each of these sailors would carry for the rest of their careers? He blamed himself and no one else. He knew he had never seen a calibration performed from start to finish, why didn't he take that opportunity three years ago to do so? He was beside himself with shame. A shame that lives on to this day.

In the next few years, Mike had observed dozens of calibration evolutions as an inspector. His feelings about the events of the three years prior continued to worsen. It became clear to him that not only could he have prevented damage by observing some of the calibrations, he missed out on the opportunity to create positive value. Specifically, the "what if?" conversations that he missed the opportunity to have while watching the evolution. Conversations that could have been triggered by questions such as:

- What if the meter was only 0.5% out of calibration? What would you do? What does the technical manual require you to do?
- What if you received a response you didn't expect? Who would you talk to?
- What if you suspected, but weren't sure, that you damaged the meter during the calibration procedure?
- What if you were tasked to complete 20 calibrations in one day because someone told you that the Engineer Officer ordered them to be compete—at all costs?
- What if you were calibrating a meter and it failed? Further, this meter was so essential to the plant's' operations that the ship wouldn't be able to get underway as scheduled until the meter was fixed or replaced?
- What if your Chief put pressure on you to fabricate successful results?

We are all subject to the temptation to take the path of least resistance. Additionally, we are all born as child prodigies with a masterful ability to rationalize almost any action if given even the most paper-thin excuse. This ability, if left unchecked, has the potential to create integrity breaches that may not even be acknowledged because integrity can become relative in the blink of an eye. This may not be music to a purist's ears, but it is true.

Observation provides leaders opportunities to not only observe employee performance but it, more importantly, allows for "What if?" discussions. These discussions are the gold mines of the observation process and are the best tool available to bridge the gap between what is expected and what occurs.

3. Lead by Example

Let's shift gears and discuss non-military organizations. It might be easier to understand the significance of integrity on a Nuclear Submarine over a software sales firm. But, although the risk of physical injury may be absent, the risk of mission accomplishment remains. If you have a culture that does not embrace integrity as a core value, you may find yourself feeling lonely—very quickly. That sense of loneliness comes from the loss of faith in the motivation of your team. Integrity in your organization is, to a large degree, earned because it is a two-way street.

Can the members of the organization trust the organization to do the "right thing"?

We believe that leaders of most organizations evaluate the answer to this question but only partially. We often unwittingly fool ourselves into thinking that as long as I do the right thing **to and for** "Ted Johnson, the site administrator," I expect that he recognizes my integrity towards him and will reciprocate.

We contend that it is not nearly that simple. Ted is smarter and more observant than we realize. How about when:

- You give him a letter to put in the company's outgoing mail that is clearly a personal and not business letter?
- Ted discovers that Client B was overbilled by 2 percent last month, and you tell him "let's not do anything unless they bring it up."
- Ted overhears you talking to a client in the conference room, and you are clearly overstating the company's track record.
- Ted inputs your expense account from your most recent business trip and notices items that look slightly overstated (as often happens).
- At a staff meeting, Ted brings to light that there is a company policy that requires us to route all vendor "Terms and Conditions" sheets through legal. You respond, "We've never done that in the 10 years that I've been here, and I'm not starting now."

You have never done Ted wrong. In fact, you have treated him better than he has probably ever been treated by a boss before. His annual bonus is always healthy as is his annual merit pay increase. You are lenient about his hours and are always patient when he makes mistakes.

Do you think Ted will embody the trait of integrity for the company? Will he do the "right" thing when no one is watching? What does he think of your integrity? If the s**t hit the fan, would you trust each other unconditionally?

The General Manager of regional office of a Fortune 500 manufacturing company is the senior person at the facility and runs a weekly staff meeting. He is relatively new, less than a year in the company and the position. The office was founded by the Flintstones and the average tenure of the employees is 300 years.

Earlier in the week, the GM was reviewing a purchase order process that didn't look correct. He reviewed the company policy, reviewed the policy with the procurement officer on the staff, and after some back and forth, she agreed that in order to be in compliance with company policy we should be doing "X" not "Y."

During the staff meeting, the Procurement Officer shares a few quasi-important matters regarding receiving and inventory. Then she turns to the General Manager and says, "Oh, and I changed the procurement process to "X" as you wanted it."

Press pause on the remote. Timeout.

Here is a perfect, but slightly camouflaged, moment for the General Manager to demonstrate his personal commitment and expectation of integrity. Unfortunately, we believe that most leaders would miss this opportunity by acknowledging the Procurement Officer's report about switching from "Y" to "X."

The organization's integrity is the protective roof to it's culture. It is not uncommon for a building to have immaculate and spectacularly designed entrances, walls, ceilings, staircases and floor tiles. If this is your building (culture), it would be reasonable to congratulate yourself for owning such a beautiful building. Recall, the roof is the least visible part of the building. Have you inspected the roof recently? What does it look like? How strong is it?

Although the roof may be out of sight and, therefore, out of mind, weak roofs will find ways to communicate with us, but these ways are usually very subtle (small leaks and creaking noises), and unless we are actively looking for them, we will miss them. If those signs are missed, the roof will eventually fail, and then no one will be able to look the other way during that event.

Let's go back to the staff meeting.

GM: *Thank you, Cheryl. But let's make something clear. We didn't change to "X" because that is what I wanted. We changed because when we reviewed the company policy, we both agreed that "X" is correct.*

Cheryl: *Right. I'm just saying that you wanted me to change it, so I did.*

(GM recognizes that this is an opportunity to seize)

GM: *Okay. Can I have everyone's attention for a moment? What I am about to say is very important.*

Room quiets down and except for the uncomfortable sound of sideways glances.

GM: *We don't do things here because that's the way "I" want it. We do things because they are the right things to do. If you ever find yourself saying, "We do this because that's the way Jim wants it done." I want you to consider that a red flag. And when you see a red flag, you stop. In this case, Cheryl shifted to "X" because we reviewed the company policy, and that is what it requires us to do. Admittedly, not everything has a policy and there may be times that we change the way we operate. I suppose sometimes it will be because that's the way "I" want it. Examples of that might include: how we run this staff meeting, how often we have this meeting, what operational events do I want to be informed of. However, when it comes to operating the business, we do the right things because they are the right things to do.*

This will not always be easy; let's not kid ourselves. However, if you are not sure what the "right" thing to do is, let's talk about it. We will involve whoever needs to be involved so that we can make the decision thoughtfully, deliberately, and collaboratively so that we do our best to get to the "right thing."

Convincing people that we (we = the organization) really do mean that we do the "right thing because it's the right thing to do" is very challenging. In part, because it is cliché and employees have likely heard this countless times before, perhaps in organizations that displayed no evidence of meaning it. Getting this message across successfully is like chopping down a tree blindfolded. You only have two choices. Place the ax on the ground and walk away or keep swinging. Worse, when it comes to establishing an organizational culture of integrity, every felled tree is replaced by another fully erect tree—new employee. There is no "arrival" to organizational integrity.

It is possible that this discussion with the staff had no lasting impact on the manner in which the staff conducts its day-to-day work. Showers have no lasting impact on our cleanliness, which is why we take them daily.

CHAPTER 7

CRITIQUES

noun cri·tique \krə-ˈtēk, kri-

*Definition : a careful judgment in which one gives an
opinion about the good and bad parts of something
(such as a piece of writing or a work of art)*

The United States Nuclear Submarine Force uses the dictionary concept of the word *critique* to launch a powerful process
with mystical powers that are guaranteed to instantly and positively transform an organization by reducing all of its obstacles
to paper-thin annoyances. Following the steps of the critique
process in this chapter will instantly raise an organization's performance to historical heights and create peace and harmony
throughout the organization.

Wouldn't that be magnificent--a checklist that traps the lightening of organizational performance success in a bottle? As
you know, there is no such a tool.

However, the Nuclear Submarine critique process is the most
formal representation of its culture of brutal self-assessment.
Without this culture, the five pillars discussed in the previous
chapters cannot be built to withstand adversity or challenge.
This culture of self-assessment provides the foundation for
these pillars, without which the pillars will be fleeting at best
and illusory at worst. This culture is inscribed on the DNA of

the Nuclear Submarine Force and is largely responsible for its unparalleled track record of safety and achievement.

Fortunately, successfully implementing this process does not require a military environment. It is teachable but only to organizations that are ready to admit that the manner that they approach mistakes and problems is often cursory and generally not effective. The challenge is that these cursory solutions are often overseen by a separate group devoted to "human performance" or "continuous process improvement." A submarine crew doesn't have that group because the process of continuous improvement is embedded into each member of the team, especially the senior leadership. Certainly, each organization is unique, so we are not suggesting that the "human performance" or "continuous improvement" groups (usually living under the HR umbrella) are not adding value. However, in many cases, their observations and conclusions are not directly integrated with the operations, sales, or service management. In these cases, the organizations are receiving next to zero, or even negative, return on investment.

One of the more maddening responses we have received when we discuss various leadership tools of the Nuclear Submarine community sounds something like this: "In the military, these types of processes work because people must do what they are told." It is true that in the short run, sailors will do what they are told. But, in the long run, sailors are human and will eventually shift from engaged to compliant and begin functioning at a minimal level of performance. Culture is not about doing work; it is about *how* people do their work. Achieving and maintaining a culture of hard work on a deployed Nuclear Submarine is a leadership challenge that has few parallels outside of the submarine. When 130 people live in a suffocatingly close environment, there are few incentives in a manager's toolbox outside of leadership -- promotions, bonuses, and time off are generally not available to the Commanding Officer or any leader on a deployed submarine. In fact, after a few months at sea, the maintenance of this culture is not only a tool of continuous improvement, it is the glue that holds the pillars together, directly impacting safety and operational performance.

Further, this culture of brutally honest self-assessment, whether it occurs in a blink of an eye in an individual's mind or through a critique that takes several days to complete, is one that the leadership must devote as much energy to maintain as it did to achieve. This staying power falls on the Commanding Officer's shoulders as the culture flows directly from him. This "trickle down" culture occurs within a chain of command of about 120 people. We recommend that this number is used as an upper limit in an organization where the establishment of a self-assessment culture is put on the shoulders of an individual. For organizations with thousands of people in many different locations (such as the Navy's Nuclear Submarine Force), the structure must be evaluated to identify what levels of management will require the most training.

The critique process is a formal process that is designed to identify the **facts** surrounding an event, the specific **problems** that led up to or were part of the overall incident, the reasons (**root causes**) these problems occurred, and the short-term and long-term **corrective actions** that will minimize the probability of this event occurring throughout the organization in the future. Amazingly, we have yet to observe a critique that did not reveal previously unknown problems about an organization's operations, performance, structure, or processes. You will be amazed to see what this critique process reveals--even within the critiques of the most (seemingly) benign and/or straightforward events.

We want to warn you: The critique process is humbling for an organization's leaders. It illustrates the "be careful what you wish for" adage. When you embrace a culture that is willing and able to perform honest self-assessments, you should expect the results to be shocking. It will remind you of the TV show *Undercover Boss* but in a less tearful and individual-centric basis. The critique process uncovers some shocking truths. Leaders are often quick to pat themselves on the back and shower themselves with praise (typically, and hopefully internally) for their inspirational, effective, and unique brand of leadership. They often cherry-pick data to support their performance assessment. The critique process strips the leader of these fabricated notions as it reveals the truth.

In the hit TV drama *Gotham*, a young Bruce Wayne (future Batman) finds a letter written to him by his deceased father. In the letter, his father warns that one can have "happiness or the truth, but not both." Although this adage is a bit melodramatic to apply to the critique process, we are going to do it anyway. The happiness of self-praise for your successes combined with the tendency to turn a blind eye to the bad stands in stark contrast to squaring your shoulders to the mirror of the organization to demonstrate your willingness to see the organization for what it is.

To an extent, we must be willing to embrace the spirit of the great Albert Einstein when he allegedly quipped, "A true genius admits that he knows nothing." The critique process asks you to set aside egos, preconceived notions, and the tendency to blame mistakes on uncontrollable external events or on a simple misstep made by one individual--as you will see, this is rarely the case. This request to be open minded is an organizational hat tip to the psychological phenomena of the external bias--the tendency to overestimate the impact of uncontrollable external factors on your own failures compared to the failures of others.

THE MECHANICS[1]

In this chapter, we will teach you the rules of chess, keeping in mind that there is a difference between people who play chess and chess players. Those who play chess need to know the rules and some basic strategies. The chess player deeply considers the various strategies, pondering all the if/then/else tactics that support these strategies. They then practice these strategies and their tactics against the best competition they can find. We can teach you how to play chess, only you can become a chess player for your organization.

The critique process is composed of three logical stages. Here is an outline of the steps that we will be covering in this chapter

Pre-Critique

- Decide: To critique or not to critique?
- If yes: Assign someone to assemble a timeline surrounding the event.
- Create the timeline.
- Identify the people who should attend the critique.
- Determine who should lead the critique.
- Schedule the critique as soon as possible.

---

1   At the end of the chapter is critique template to allow you follow along more effectively with the process that follows.

THE CRITIQUE

- Lay down the groundwork.
- Review the facts and the timeline.
- Identify problems (actions that led to the event that violated standards of safety, technical instructions, or organizational standards)
- Determine the root causes of these problems. (The most challenging and important step)
- Assign corrective actions to the problems that will address the root causes. These are typically divided into short and long-term corrective actions.

POST- CRITIQUE

- Routing a smoothed copy of the critique report through the chain of command (see template in back of this chapter)
- Track the completion of corrective actions

PRE-CRITIQUE

The Director of Marketing forgets his wallet at home. He is already halfway to work when he realizes this so he goes back home to retrieve his wallet. As a result, he is five minutes late to the morning staff meeting. His company is not going to launch a formal critique process. (Although, the Director of Marketing will perform a mini-critique in his mind to include corrective actions. "I'm never leaving my wallet in the kitchen again!")

If the Control Room Operator in a power plant violates the procedure for starting a turbine generator, which results in physical damage to the turbine blades and lost revenue for the company, the officer in charge will launch a very formal critique process.

Where is the line between these two scenarios regarding the activation of the critique process? There is no exact answer. You will derive your own threshold, which will be based in large part on your specific organization, industry, and personality. The critique process begins at the moment you ask this question.

Decide: To Critique or Not to Critique?

Some ideas to consider when developing your threshold include the following:

- What were the consequences of the mistake?
- What were the potential consequences of the mistake?
- Is this issue a recurring one?
- How many people does the issue involve? (The more people involved the more likely a critique will be necessary.)
- How confident are you that you know exactly what happened and why? (Be very careful here.)

An effective way to determine the answer to the last question is to ask the supervisor responsible the who, what, where, why, and how of the scenario. If the answers to any of these questions are unknown, you still have some fact-finding to progress through before you can decide. If you observe hesitation or uncertainty, don't ignore it.

On Nuclear Submarines, we had a middle ground process (between doing nothing and activating a critique) called a fact-finding meeting. This involves fewer people and is less formal than a critique. The fact-finding meeting delivers exactly what its name implies--facts. The leader must leave this meeting feeling assured that there isn't a deeper layer to the problem that needs to be explored.

This process is not perfect and is without a doubt an art. We have seen examples of leaders who fall at the far end of each spectrum and hold a critique for everything or nothing. Trust that you can find the correct middle ground for your organization, knowing that that middle ground will continue to shift slightly in both directions on the spectrum based on the contextual variables in your organization's world.

Assign Someone to Assemble a Timeline Surrounding the Event

If you decide to proceed, the effectiveness of the critique process is dependent upon the accuracy of the timeline. This is a very simple case of "garbage in results in garbage out." Many well-intended and executed critiques are a waste of resources, time, and money because they were based on an inaccurate or sloppy timeline.

The timeline is exactly what it sounds like--a timeline of events that include the date and time of the circumstances and actions that led up to the event, the event itself, and actions after the event, up to and including the situation being brought to a steady state. As you can imagine, there is a spectrum of detail in each timeline. Some events need minute-by-minute details, while others need no more than day-by-day details. In any event, this is not an easy task--ever.

A common mistake is to assign the wrong person to create the timeline. There are four criteria to use when selecting the best person to develop the timeline:

1. The person was not directly involved in the untoward event but has knowledge of the event's occurrence.

2. The person is familiar with the operations and vernacular of departments involved. As an example, a person from human resources is not likely to generate an accurate timeline regarding a failure on the production line of a manufacturing company.

3. The person must demonstrate the highest levels of "Questioning Attitude" and "Integrity" as described in Chapters Three and Five.

4. The talent and seniority level of this person is likely higher than you might initially assume, especially if you are in the initial stages of introducing this process to your organization. This is not an administrative task.

Create the Timeline

Developing the timeline starts with asking all those who had a role in the event to write a statement that describes the events as they can best recollect including their actions and the actions of others that they observed. This statement should also include the best estimation of dates and times. This is likely to be an interactive process. It is our experience that most statements will have to be redone because the initial statements are too vague or leave out vital pieces of information. There are reasons for this that should be considered before starting this process. Many people will interpret "writing a statement" as part of an inquisition.

The person at the helm of the timeline will then have a jigsaw puzzle on his hands. He may be holding between two and twenty statements that now have to be merged to create one timeline. With each statement collected, merging them becomes exponentially more difficult. The person tackling this task will have to discuss the statements with their authors to clear up discrepancies. For example: John says that he came back from lunch at 12:30 and then went to Bob's office to discuss Project X. However, Sandy's statement says that John came back from lunch at 12:00 and immediately began working on Project Y (with no reference to his trip to Bob's office). These discrepancies are rarely a result of an attempt to hide the truth; they are the result of the human brain. We all see and remember the same events differently and the size of that difference increases with time. This is why it is imperative to get the timeline created without delay.

The goal is to create a timeline that is the most reasonable set and sequence of events that can be obtained without channeling the investigative prowess of Sherlock Holmes. This is not a witch hunt and a sufficient amount of effort from the organization's leadership will be required to get this message across. The message is simple--we want you tell us what happened and when. There is no need or benefit for a statement to include speculations about why something happened or to be defensive.

When you receive the completed timeline from your designated Sherlock, expect to send him or her out for a second attempt. It will take time for the organization to understand what a good timeline looks like. Without having the benefit of attending or leading critiques your request for a detailed timeline may seem elusive or even confusing. Expect this response, and do your best to nurture your Sherlock during this process such that he could soon train someone else in a similar fashion.

In the Submarine world, we would often have a working time-line within a few hours of even the most complex events. This ability was partly due to the ingrained culture and the collective understanding of how the critique process worked. However, we also had a captive audience. When you're hundreds of feet deep in the middle of the Pacific Ocean, there are no fathers trying to get out the door to see their sons' soccer games. Also, and more significantly, we had no human resources or labor union gauntlet to run through. We recommend discussing this process with both groups before you have an event to critique. This will allow for questions and concerns to be discussed without slamming on the brakes of a critique that you need to perform immediately.

Identify the People Who Should Attend the Critique

If in doubt about whether someone should be on the critique attendance list, the answer is "yes." This is especially true as you are introducing the process to your organization. Over time, you can back off, but in the beginning, it is better to have too many in attendance than too few. However, keep in mind that the larger the group, and more specifically the larger the percentage of senior personnel in attendance, junior workers and more shy types may be less likely to contribute when you need them to--such as speaking up, unsolicited, to raise an inaccuracy of the timeline or a salient point that was missed when discussing the how/why of a particular action.

In determining who should attend a critique, there is no one answer that will work across any range of variables such as the magnitude of the event, the number of people involved, the uncertainty surrounding the events, and your personal preference. The general advice is to include the following people:

- The individuals who were directly involved in the event
- The individuals who witnessed or were affected by the event
- The supervisors of these people
- The critique leader
- The senior-most person the organization can devote to the process

Additionally, sometimes a critique can be used as a training exercise by inviting a handful of people who have no involvement with the events at all. As an example, two supervisors from the marketing department might be invited to observe an Operations Department critique. There is obvious value in doing this, but the technique should be used sparingly and selectively as you do not want the members involved in the event to feel like they are being watched like animals at a zoo.

When informing people of their selection for the critique group, you can be sure that the following scenario will happen: One of the supervisors will claim that he can represent one or more people at the critique. The reasoning will vary slightly but it will resemble one or more of the follow rationales: "They are in the middle of a job that is of critical importance," "So-and-so requested that day off," "I have spoken to them at length, and I will be able to share the information more clearly than they would."

In evaluating your answer, we recommend you weigh the pros and the cons, and then say "No!" There are exceptions to every rule, except this rule--usually. Clear as mud? While, we can't say there will never be a scenario in which it will be acceptable for a supervisor to replace a designated group member, it is challenging to think of one. Neglecting to bring in the people who were closest to the problem degrades the critique's effectiveness. In fact, it may actually be a red flag. We are not dismissing the possibility of the request being a sincere one, but you shouldn't dismiss the possibility that the request is an attempt to hide something. This is so challenging to say because we really do believe that people are generally well-intentioned and having a predisposition to distrust is cancerous. However, we've seen this pattern enough times that we'd be remiss not to mention it.

If you do entertain these requests when they come up, and they will, it is a mistake and a horribly common one to rationalize them away as the cost of doing business--"You can't always take away a person from the job of making money for the company for an administrative meeting." You are steering into shallow waters if you accept this. The critique process must evaluate the event with information that is least susceptible to the telephone game phenomena (the kindergarten game that proves people can't reliably pass along a one-sentence message). As soon as you remove a person closest to the event from the critique group, you can guarantee that a substantial piece of information will remain unknown. We have participated or observed hundreds of critiques and in every single one, information came to light that no one could have predicted. More often than not, this information came from the most junior person in the room.

The chain of command concept is sacred in the military, but we've observed it to be an important part of any civilian organization as well. There is an element of the critique process that must be handled with care in regards to the chain of command. We are intentionally placing people in various levels of the chain in the same group and giving each person the ability and obligation to speak freely. In an ideal world there are no negative consequences to sharing information in an organization that values integrity and transparency. However, the world is not ideal. Some people will be reluctant to share information during a critique. There are several reasons why and among the worst is fear of retribution from their immediate supervisor. If this is true, you have a problem that needs to be addressed aggressively; however, a false accusation can be as damaging. Be on the lookout for other indications that might substantiate this.

On the flip side, the critique leader must also be sensitive to the unique dynamic that develops in a critique and how it relates to the chain command. For example, if a junior mechanic shares information to the critique leader about how or why he performed a particular function, and this is followed by a barrage of pointed questions that feels aggressive or accusatory, that junior mechanic may think twice about opening his mouth during a critique in the future. Who wants to put the boss on the spot? The moral of the story is that the critique process does mishmash the chain of command, but it is done for the good of the organization. The critique leader must find a way to balance this unique situation or future critiques could be extended versions of "name, rank, and serial number." We've seen that happen, and it is not pretty. Additionally, it is an indication that an organization's culture is shifting in the wrong direction.

Determine Who Should Lead the Critique

Let's start by stating who should *not* lead the critique--the senior person in the room. On a Nuclear Submarine, the Commanding Officer would attend most critiques but rarely lead any of them. It would be most common for the respective department head to do so. However, sometimes the event resulted in consequences so dire (a death, broken equipment that shuts down plant operations, the loss of the company's largest customer) that it is necessary for the senior person to lead the critique.

Not utilizing the most senior person may seem counter-intuitive since we've emphasized the importance of the direct involvement of the most senior people of the organization in this cultural-shift process. That is a reasonable reaction. In fact, the most senior person is most likely to be the most skilled at leading a critique. However, the benefit that this person can provide as an observer outweighs the incremental difference in leading the mechanics of the critique. There are many reasons for this, most of which can be summarized by making efforts to avoid "poisoning the well." The group will, often unknowingly, synchronize their beliefs with the most senior person in the room. We want to avoid "group think" because it has the potential to nullify the benefits that we are trying to achieve. (Think back to the previous chapters on Questioning Attitude and Watchteam Backup.)

Here are our guidelines to the most senior person in the room:

1. Keep your mouth shut. For many senior leaders, keeping their mouth shut is more difficult than holding their breath.

2. Keep your poker face on. When there is turbulence aboard a commercial flight, where do most people's eyes go immediately? To the flight attendants. When turbulent information is revealed, all eyes will look to the senior person for cues on how to feel about the severity of the situation. Do not allow this to happen.

3. Be the safety net to the critique leader's tightrope walk. If the leader is losing control, help her regain it. The most common scenario will be allowing the group to proceed down a rabbit hole (a topic that pops up and is worth pursuing but is unrelated to the events that are

being critiqued).

4. Press pause or rewind when the veracity of the critique is at risk. If a critical piece of information was glossed over, was not obtained, or was mishandled, speak up to ensure it is addressed.

5. Observe the tone of the room. Did it change at a particular moment? Was there an individual who seemed particularly disengaged, uncomfortable, frustrated? These intangibles are routinely missed by organizations but the observed emotions, collectively and individually, are a source of knowledge about the organization's culture that is limitless. Who better to evaluate this than the you?

6. When the critique is over, thank the group. Regardless of how you may be feeling at the time, which predictably may be a combination of frustration, anger, disbelief, and humility, suck it up. Provide the group members a sincere gesture of gratitude for their participation, their honesty, and their effort. Take a moment to remind the group why we do these critiques. This is your opportunity to influence their emotion at the dinner table with their families.

Schedule the Critique as Soon as Possible

It is vitally important to conduct the critique as soon as possible for two important reasons:

1. When you prioritize the critique above other organizational activities, you send a message that no words could. This process of self-reflection and continuous improvement is not the new

corporate lingo that so often dies on the vine or survives as a self-licking ice-cream cone. Instead, this is at the core of how we do business. It is not something we are doing; it is what we do.

2. With each minute that passes, people's recollection of the events diminishes. The details are the first to go--you can't let that happen. Further, the recollections of two people observing the exact same event with the exact same vantage point will vary only hours after an event. These differences will continue to diverge as time passes. (Many books would reference a study to substantiate this phenomena, we will instead refer you to the 456 episodes of "Law & Order.") There are few moments in a critique that are more frustrating than losing a crucial piece of information. The words, "I don't remember" is a damaging dagger for which there is no defense.

On a Nuclear Submarine, it is common to hold the critique the same day of the event. This often causes the stress levels to rise and the sleep hours to drop for the people involved. However, in the high-risk environment of a Nuclear Submarine, the risk of a recurrence of the untoward event (or a related one with shared root causes) is too high to accept a delay. This would be embarrassing at best and life-threatening at worst.

THE CRITIQUE

You are the critique leader. You've completed your preparations, and now the team is assembled and ready to go. The critique leader can now jump right into a discussion of the timeline events. Right?

Not so fast.

We know just how tempting it is to dive right into the meat of the critique--launching right into the timeline and the events. You mean business, you are respectful of people's time, and you are prepared to dig as deep as possible into the details of the event to maximize the benefit to your organization.

We get it. But first you need you to slow down and take a deep breath.

Lay Down the Groundwork

Before you begin a discussion of the actual event, be sure to attend to three critical items:

1. Verify attendance.

2. Ensure the participants are comfortable and seated so that each person can hear and be heard.

3. Review the rules.

1. Verify attendance. At least one person who was identified as a participant will not be there. This is a statement fact. Invariably, someone will be on vacation climbing Mount Everest, out sick with the Ebola virus, single-handedly repairing the company's most vital piece of equipment, or meeting with a potential client whose business would triple the company's top line. This is a test, and all eyes are on you to see how you will handle it.

This scenario is both inevitable and consequential so it is worth thinking through your response beforehand. You have a few options. Let's look at the two ends of the option spectrum.

At one end of the spectrum is to take a rock hard stance. You want the message to ring loud and clear and in no uncertain terms: if we have a critique, everyone scheduled to be here is here. Period.

Here is a true story that defines this end of the spectrum:

When a submarine is in port (home) the time the crew can spend with friends and family is precious, deserved, and overdue. This is especially true for members of the Engineering Department who are often required to work long hours during maintenance periods, which can only be performed when the nuclear reactor is shutdown. Often the families of the Engineering Department members have a difficult time understanding and accepting this harsh reality. After all, in some cases their loved one was away for months at a time. Therefore, the time sailors can spend with their families is precious, to say the least.

In this instance, the submarine had recently completed a two-month underway period and was in port and conducting a particularly challenging maintenance activity period. On a Friday afternoon, a relatively simple maintenance item was performed incorrectly largely due to a lack of procedural compliance. There was no damage or injury; however, the Engineer deemed that a critique was necessary. The Commanding Officer agreed. The critique was scheduled for 0700 (7 a.m.) on Saturday morning and was to include 10 members of the 55-person department, including three supervisors.

At 0700, the group was settled in the wardroom and prepared for the critique. However, there were 9 people--not 10. The Chief of the absent mechanic (who we will call Petty Officer Silver) contacted him. Petty Officer Silver slept through his alarm clock but said he would be on the boat by 0730.

The question now is: How do we handle the absence of a person scheduled to attend a critique? Reasonable options at this point may include: Start the critique without Petty Officer Silver (his role in the event was very minor), wait for him, reschedule the critique, or . . . Recall the entire Engineering Department to the boat. That's right. The Commanding Officer ordered the Engineer to recall the entire Engineering Department to the boat. The critique would be rescheduled for tomorrow (Sunday) but no one was going to leave until we had the entire Engineering Department on the boat.

Why?

To make a strong statement that our submarine does not take attendance at critiques lightly. The Commanding Officer wanted to draw the attention of the entire department to Petty Officer Silver's mistake to prevent anyone from making the same mistake. Everyone was stunned. This directive required calling people off of leave (vacation), taking them away from the rare weekend family activities, and disrupting all weekend plans of the crew. We were on the submarine until 9 p.m. that night.

There was not much evidence that this decision improved the culture, motivation, or performance of the crew. However, no one would ever be able to question the Commanding Officer's commitment to the sanctity of the critique process. This Commanding Officer was later selected to be an Admiral in the United States Navy.

This is one side of the spectrum for you to consider.

The other side of the spectrum is to accept that assembling a group of people, sometimes from different departments, places a significant stress on the organization. Let's accept what we have, move on, and make the most of it.

Somewhere between these two extremes is your answer and no one can define this for you. Bear in mind that how you handle this situation will send a message to the group that will trump any verbal message.

"What you do speaks so loudly, I can't hear what you are saying"
- Ralph Waldo Emerson

2. Review the rules.

Quiz: If your next critique is with the exact same people two hours after this one ends, will you review the rules of the critique before you start?

You may write the rules and read them each time, or you may choose to speak off the cuff and paraphrase. In either case, write down the rules for your own benefit. This exercise will remind you of your promise to the group. We can assure you that making good on your promise will not be easy. Here is an example:

*"We are here to critique an event (insert event). We are committed to continuous improvement because there is always room for us strengthen the organization and ourselves as individuals. We may not like it, but mistakes happen. We want to learn as much as we can from them and take the necessary action to prevent a recurrence throughout the organization. We are **not** here to point fingers or assign blame, we are here to take an honest look at ourselves and*

the manner in which we operate. Accomplishing this requires open, honest, and respectful dialogue throughout this process. Let's take a moment to put our egos aside and review the event objectively and not defensively.

Any questions before we begin?"

Review the Facts and Modify the Timeline

This process is straightforward but cannot be rushed. Do not assume that the timeline assembled by your Sherlock is without error. In fact, it is better to assume the opposite and draw out those errors during the first stage of the critique. You can not begin to discuss problems and root causes effectively until the facts of the event are understood and agreed upon. Of course, there will be times when there will be a dispute about a fact that will not be resolved. The leader does his best to mediate this disagreement until it is resolved.

Everyone in the room should be given the time to review the timeline (usually through printed handouts). This process will typically take 5 -- 10 minutes. Ask people to hold their comments until everyone has completed the review. Ask everyone to look up when he or she is done reviewing the timeline. At this point, open the floor to recommended changes to the timeline. The timeline should also be displayed on a screen with a person assigned to edit the displayed version during the critique.

When establishing the timeline, resist the temptation to dive into problem identification phase. Or even worse, avoid turning this into an inquisition phase (which is never appropriate) that sounds like this: Supervisor X to Worker Y: "Why the hell would you do that? I've told you a million times; in fact, I even reminded you that morning." The leader of the critique must squash that type of exchange immediately. Be prepared because it will happen. Do not allow discussions other than statements of fact and the time they occurred. (Off-track example: "Why do we perform b with c? I remember seven years ago when we used to . . .")

On the other hand, resist the temptation (which is more common) to assume the timeline was written by hand of God. Be prepared to modify the timeline through a civil, respectful, and relevant conversation. Perfection is not attainable. At some point, you must decide to move forward with what you have, even if disagreements are still unresolved.

With a timeline projected onto a large screen, the critique leader should now review the timeline by reading aloud each event, slowly and deliberately. As he goes through each event, the critique leader (and others) should feel free to ask questions to the individuals involved. What type of questions? Questions--without criticism--that get to the heart of why someone did something. This is critical to an effective critique but also extremely challenging. You will encounter dialogues such as this:

> *You: You intentionally bent the flux capacitor before you installed it?*
>
> *Other Person: Yes*
>
> *You: Why did you do that?*
>
> *Other Person: I thought that I was supposed to.*
>
> *You: Do you remember why you thought that?*
>
> *Other Person: No. (awkward silence) I'm pretty sure that's*

what I learned during my training.

You: Did you refer to the procedure during this process?

Other Person: I didn't.

You: Was there a pre-evolution brief?

Plant Supervisor: Yes, but "other person" wasn't in attendance.

You: Why not?

Plant Supervisor: He was at lunch.

As difficult as it will be, the critique leader cannot break out into beast mode. *"Why in the world would you hold a pre-evolution brief without the primary mechanic in attendance?! What is wrong with you? We've been over this a thousand times!"* If you do that, you are violating the promise you made to the group at the onset. If you do that, you assure that others will clam up as the critique proceeds.

However, that doesn't mean you shouldn't feel comfortable asking questions until you fully understand what happened and why. In fact, you must do so.

Keep your questions respectful and unemotional. Make it obvious to the group that the only reason for the questions is to understand the events. If the action is a "problem," (organizationally defined--examples include safety violations, technical errors (procedural, best practices), poor communication, violation of organization standards), direct the typist to place a numbered "P" for problem in the margin next to the event on the timeline (P1, P2, etc.). Other members of the critique should feel empowered to identify an action as a problem but the critique leader must maintain control of the conversations at all times. Often, supervisors may start down a finger-pointing path with the members of their department. Squash this quickly and respectfully.

Asking the right questions, in the right manner, is the most challenging part of the critique. You need to dig until you truly understand what happened and why. The digging is performed through your ability to ask probing, and sometimes uncomfortable questions. As an unofficial guideline, if you did not uncover something that was completely unexpected, you probably didn't ask enough questions.

However, keep in mind the post-critique impact on the organization's culture and moral. Ultimately, an organization wants the thought process represented by the critique to infiltrate every layer of the chain of command. If the people who participated in the critique feel like they were publicly humiliated or treated disrespectfully, word will quickly spread that critiques are a nightmare and should be avoided at all costs. Perception is reality; therefore, do your best to monitor and control the perception.

Identify Problems

As the timeline is reviewed, problems are annotated on the timeline. Control the number of problems. Without a filter, the timeline may have 15 problems. This is too many to handle for one critique. Often you may have the same problem identified multiple times. When this occurs, consolidate them. For example, if P2, P4, P8, P11, P15 are all "poor communication," than change each of them to P2--"poor communication." Keep the problems focused on the ones that contributed to the event that you are critiquing. There is no doubt that other organizational problems will be uncover. They need to be addressed (an added benefit of the critique process), but not now.

As you can see in the template located in the end of this chapter, the problems are first identified on the timeline with a "Px" notation without any elaboration other than a note that indicates "a problem occurred at this step." When we transfer that problem to the next section of the critique report, the problem should then be defined with a sentence or two. Recall that the ultimate goal of the critique process is to improve the performance of the entire organization. Therefore, the problem statements are often best expressed in more general terms than you may initially be inclined to describe them. For example, assume that the frontline supervisor and the project manager had a miscommunication which resulted in a four-hour delay on a manufacturing line that day. (Further assume that this delay is not the event this is being critiqued but instead contributed to the department's failure to meet its weekly production quota).

It is not uncommon for us to read the associated problem as, *"John B (the mechanical piping frontline supervisor) failed to provide Allen R (the project manager) enough information about the unexpected reduced manpower available to support Tuesday's production. If John B had communicated the manning shortfall to Allen R earlier in the day, Allen could have reallocated manpower resources or modified production line utilization to reduce the negative impact on the department's daily production."*

This statement may be true, but it is a poorly written problem statement. Its length and specificity undermine its effectiveness. Saying that specificity is not desirable may seem counterintuitive. However, recall that a goal of the critique process is to improve the performance of the entire organization without giving anyone the impression that the process is a blame game. An indication that your critique process is gaining traction is when a critique report about an event in the maintenance department results in improved practices in the organization's marketing department.

The specificity of the example problem statement is flawed because it leaves the door open for the individuals involved to carry on a dispute of the specifics. The specifics no longer matter--"a miscommunication occurred." That is a statement of fact and it is a problem. Also, if the problem statement is too specific, people in the organization are more likely to dismiss the problem as one that "does not apply to me because I don't work on the production floor and don't have those types of problems."

Generally, (although the specifics of the actual event may modify this slightly) you should strive for a problem statement that is worded closer to: "Poor communication between supervisors during the exchange of daily and routine information." This problem statement is true, specific enough to capture the essence of the problem, and general enough to encourage supervisors in other departments to consider the information applicable to their daily routine as well.

Another contrast between these two versions is that the second one does not elaborate on the consequence of the problem. You should already know what the consequence was. In this case, "the department failed to meet its weekly production goal." There is no need for the problem statement to restate the event that is being critiqued. This is not a trivial suggestion. Critiques will, in time, allow an organization to assess its strengths and weakness and also to evaluate the effectiveness of critique corrective actions. For example, if asked, "What percentage of our critiques in 2016 contained a *miscommunication* as a contributing problem?" the simplified problem statement allows for more effective and efficient trend analysis and data mining.

Assign Root Causes

Buckle up. Almost every organization has a process the resembles the one we have described in this chapter. In fact, every human brain has a process the resembles the one we just described. Determining the root cause is the where the money is. This is the point in problem analysis where most organizations fail. The worst part of this failure is that it goes unnoticed. These organizations proudly declare that they have an RCA (root cause analysis) process that evaluates, records, and shares the root cause for every safety violation and operational error. In many cases the process to which they are referring is solely an administrative drill. In fact, we've witnessed companies that mandate a quota system for reporting "near misses." As a result, there is a program administrator somewhere in the company with airtight evidence of the company's commitment to continuous improvement through self-evaluation and root cause analysis.

The overwhelming majority of these programs add very little, if any, value to the organization. There are very few sweeping generalizations about organizational behavior that we are willing to make; this is one of them.

If we held a critique about why an organization's post-mistake process is ineffective, "inadequate root cause analysis" would be the root cause of the majority of them. The rest would be "inadequate corrective actions" (which we address next). In either case, the process is administratively laborious not unlike digging holes and then filling them back in--a lot of energy is expended with nothing to show for it.

After reading to this point, you may have decided that we are condescending jerks. Good. This is intentional. If we had the ability to scream one organizational management message from the mountaintops to every leader and especially the CEO or president, it would be: "You suck at identifying the root causes of your problems, which is why you repeat them." A mitigating factor to this deficiency is that getting to the root cause of an undesired event or outcome is never as obvious as it seems at first glance--nor should it be. Consider encylopedia. com's definition of the word *root* in a botanical context: *"the part of a plant that attaches it to the ground or to a support, **typically underground, conveying water and nourishment to the rest of the plant** via numerous branches and fibers."* The root is the reason a plant grows or wilts, and it is not visible without digging deeply underground. The older or bigger the plant, the deeper you have to dig to find the root. The same is true with organizations.

The easiest way to determine if an organization is off-track on its root analysis process is to review a list of them. The dead giveaway is when the majority of the root causes are placed on the individual actions of personnel. The root cause is rarely, VERY rarely, an individual error. The challenge is to rewire our brains to place more responsibility on the organizational processes and systems as well as on the talent of the supervisors.

Let's demonstrate how our brain performs its own internal critique in response to an undesirable event.

> *Event: I slipped and fell on an icy sidewalk and bruised my butt.*

> *Problem: I wore shoes with slick soles.*

Root Cause: I didn't realize how icy the sidewalks were.

Short Term Corrective Action: Get up.

Long Term Corrective Action: Pay more attention to the weather in the future and wear more appropriate footwear.

This process covers all of the elements of the critique process (except the timeline) and occurs lightening fast. In fact, this critique process was probably over before your butt hit the ground. The evaluation and plan for the future are all reasonable conclusions. However, this thought pattern does not translate well when evaluating professional untoward events.

Consider an analogous event at work and the consequences of using the same thought pattern.

Event: A delivery truck broke down and a time-sensitive delivery to our most valuable customer was not delivered on time.

Problem: Most important delivery was placed in our least reliable truck.

Root Cause: Dispatcher did not realize any of the day's shipments were time-sensitive.

Short Term Corrective Action: Reroute another truck to pick up and deliver order.

Long Term Corrective Action: The shipment dispatcher will pay more attention to time-sensitive shipments and place them in the most reliable trucks.

Our brains love patterns. They spend their days (and nights) searching for patterns. The more patterns they can identify, the more situations we are equipped to deal with. However, our brains are naturally wired to protect us from being eaten by a lion--not to lead a logistics business unit. Therefore, we must resist the temptation to use the pattern that our brain is trying to push on us. This is a lot easier said than done, especially if you are not trying—and therein lies the rub.

Without having even explored the details of this hypothetical and simple example, we know that the identified root cause was a leaf not a root. As we mentioned earlier, if the root cause is a variant of "someone made a mistake because people make mistakes," this is a red flag that the critique process is not serving the organization well. (Admittedly, sometimes the root cause is "Johnny didn't do his job," but those situations are extremely rare, and there is usually a reason Johnny didn't do his job, which is the real root cause.)

When we misidentify a leaf for a root, no sirens blare and no neon lights flash. The the most common response is the one of least resistance and the one that resembles our intuitive process of "I guess I won't do that again." The team will deliver the shipment as soon as possible and will likely extend all of the appropriate apologetic courtesies. Maybe a supervisor sits down to reprimand, counsel, or coach the dispatcher. Or maybe not. Is it fair to expect dispatchers to able to predict when a truck will break down? After these "immediate actions" are complete, we move on with life. Our inboxes our full, the dragons are queued up waiting for their chance at you, and new shiny objects glitter about the office space. We have all been members of organizations where this is precisely the form of the post-mistake response and where there was no post-mistake response. In fact, everyone reading (or writing) this book has swept a mistake under the proverbially rug. It's (usually) not criminally negligent to do, and sometimes it is required because you assess the magnitude of the error to be small, or because you are crushed under the weight of other time-sensitive material, or because your boss ordered it. (We'll come back to the last one.)

We will elaborate on root cause analysis in the next chapter, but for now, know that we contend that there are only a handful of root causes in the Nuclear Submarine world. They are:

1. Inadequate supervision

2. Inadequate training

3. Inadequate standards

4. Inadequate policy / procedure

5. Inadequate communication

We know that this list will not be exhaustive for every organization and industry. However, keeping your list compact will assist your organization's consistency. Change is challenging. Changing culture is very challenging. Changing culture without consistency is more challenging than completing a six-sided Rubik's Cube in a dark closet shared with a hungry honey badger.

Assign Corrective Actions to the Problems that will Address the Root Causes

X marks the treasure on a pirate's map. For us, X marks the corrective actions. Without corrective actions, a critique is a chat with strict guidelines and would have been more fun over beers after work. Unfortunately, many organizations dismiss the idea of learning how to effectively use corrective actions because they contend that they already do so. It has been our observation that organizations suffer from this delusion only because they view the words "corrective" and "action" to be simple ones and then extrapolate a very dangerous assumption about the challenge of assigning "corrective actions." We hope the same is not true for hang gliding.

Learning to effectively use this critique process to improve the performance of your organization is analogous to learning how to play golf. If you haven't previously been acquainted with the sport, it appears to be offensively simple.

The golf swing looks like it should be as challenging as swapping the clothes from the washer to the dryer. However, when you pick up a club for the first time and attempt to hit that small motionless ball, your humility meter pegs high. However, the sport is a black hole, and once you get "it," there's no looking back. With practice and the right coach, you will improve, but perfection is unattainable. Assigning corrective actions is like learning to putt. It appears to be the easiest and least sexy aspect of the sport. It is baffling to learn that many professional golfers spend half of their practice time putting. However, early in your golf career, you quickly realize that you can't score well in golf unless you are a good putter. Period. Similarly, your efforts to strengthen your organization's culture of brutally honest self-reflection will be in vain unless you learn the art of corrective actions.

Before delving into the mechanics of assigning corrective actions, let's start with the "6 Commandments of Corrective Actions":

1. Short-term corrective actions are materially different

than long-term corrective actions.

2. Each problem identified in the critique has a corrective action assigned to it.

3. Find your second wind before assigning long-term corrective actions.

4. Corrective actions are not punitive.

5. Corrective actions must address the root cause(s) of the problem.

6. Corrective actions without a formal training program are like new recipes without a stove.

Let's review the basics of each of the commandments.

1. Short-term corrective actions are materially different than long-term corrective actions.

Short-term corrective actions are assigned to stabilize your situation. What needs to occur in the upcoming days to prevent this mishap from recurring. Long-term corrective actions are assigned to improve the organization's future. An effective mental model to distinguish the two is to the people to which these actions are assigned. Short-term corrective actions generally apply to your organization as it is manned and organized today. Long-term corrective actions apply to your organization as it is manned and organized in the future. Long-term corrective actions can be considered long-term contributions to the organization's future success. You and your team are not likely to be the benefactors of long-term corrective action, the people who hold your position in the future are.

2. Each problem identified in the critique has a corrective action assigned to it.

During the critique, we identified "problems" as we reviewed the timeline of the event. Each of these problems were then assigned root causes. Now it is time for us to solve these problems by assigning corrective actions. In theory, if our corrective actions are exactly correct, the probability of this critiqued event from recurring would be zero. However, as some claim Yogi Berra said, "In theory, theory and practice are the same. In practice, they are different."

Typically, at least one short-term corrective actions and one long-term correction action is assigned to each problem, but this decision is contextual and is only a rule of thumb. Further, assign as few corrective actions as possible but no less than is necessary to adequately address the root causes of the problem.

Short-term corrective actions address elements of the event that eradicate cancer cells that are multiplying as we speak. The longer we wait, the bigger the tumor.

Assume that an expensive piece of equipment (a flux capacitor) was damaged during operation. One of the problems identified was that the operator was following the incorrect version of the operating procedure. Your production plant has five flux capacitors with fifteen qualified operators. You need to verify that the correct procedure is being used for all of the flux capacitors by all of the operators. This is an example of a short-term corrective action.

Long-term corrective actions are steps that we can take to prevent the cancer from coming back. If the critique identified that your organization does not have a formal process for implementing and controlling revisions to procedures, assigning the development of such a program may be an effective corrective action.

3. Find your second wind before assigning long term corrective actions.

Critiques can be time-consuming. In fact, when you first implement them using the Nuclear Submarine model, they **will** be time-consuming. As your organization becomes more adept at the process, you will become more efficient. However, whether the critique lasts an hour or five (ouch!), the process is usually mentally exhausting. Since the corrective actions are the last step, it is not unusual for a well-performed critique to be sabotaged by ineffective corrections actions because the leaders felt rushed to wrap it up (which would be a nice way of saying that the leaders allowed themselves to be mentally lazy).

The temptation to fly through the corrective actions is natural. If you've led a good critique, you will feel exhausted at this point. Therefore, protect yourself from yourself through whatever mechanism works best for you. Some leaders prefer to break for 10 minutes every hour to minimize this mental exhaustion. Others choose to break for 10 minutes prior to starting the corrective action discussion.

Short-term corrective actions may require immediate action and leave no time for a break. If you identified a root cause that could allow a similar problem to continue to occur RIGHT NOW, you must take action to reduce or eliminate that root cause through a short-term corrective action. For example, if you identify that some of your operators were trained wrong on operating procedures, you may need to place supervisors to watch over them until you can hold training.

4. Corrective actions are not punitive.

In the last chapter, we discussed Integrity. We discussed the importance of trust up, down, and across the organizational chart. You led the meeting by reading the "critique rules," which included the assurance that you do not convene critiques to point fingers or to punish. You critique so that you, as a team, continually improve your performance. Corrective actions that are punitive, even if they are thinly veiled as productive, will immediately reduce the credibility of your critique rules to "just another bunch of corporate B.S. from a suit."

Consider the operator that used the incorrect version of the flux capacitor operating procedure. For one reason or another, you are furious (on the inside) at this operator. Therefore, you assign him to handwrite the correct operating procedure by hand 10 times as a corrective action. This is punitive no matter how you spin it.

5. Corrective actions must address the root cause(s) of the problem.

This is fundamental yet tricky. We've banged the "problem – root cause – corrective action" drum, and hope you have to. Therefore, this commandment should be a slam dunk. You can't stop a problem unless you attack its root cause. It sounds easy but never is.

Assume that you identified that poor cleanliness in the flux capacitor workspace was a problem that contributed to damage. During the critique, when the time came to identify the corrective actions for this problem, your supervisors debated for awhile.

The noise sounded something like - *"He [the operator] always keeps his workspace messy . . . I have told him a thousand times . . . it's not just him, the entire flux capacitor space is cluttered . . . we are not talking about clutter . . .we are talking about cleanliness . . . clutter is part of cleanliness . . . the plant in Tulsa is much worse . . . I was in Milwaukee last week, and they keep the place spotless."*

You interject, "Enough. The root cause is that we have inadequate standards of cleanliness throughout the operating facility. It's not just the flux capacitor areas that are mess. By my standard, the entire facility is filthy."

The air is sucked out of the room. Supervisors lean back with folded arms and shoot each other sideways glances. They disagree.

Based on this information, grade the the following corrective actions (A-F).

Short-term corrective action: Lead Operator will clean his flux capacitor workspace.

Long-term corrective action: Effectively immediately, each front line supervisor will perform a daily evaluation of the cleanliness of her workspace and formally counsel any operator that leaves the workspace dirty.

We're not going to compare grades, but we hope that you agree that this is not "cover with gold stars and hang on the refrigerator" caliber. If low standards of cleanliness is the root cause than our corrective actions must address the standard.

Cleaning one-work space one time is not even nibbling at the edges of the standard. The long-term corrective action is a bit trickier. Directing supervisors to inspect the areas and to counsel non-compliant operators *sounds* reasonable. Given the limited information, we have shared about this event, the corrective action is junk. If an organizational standard is a problem, it is not reasonable to expect those who have allowed this low standard to raise the standard themselves. Modifications to standards is a Herculean task, so when you identify a root cause that calls out the organization's standards, do not expect simple or conventional corrective actions to work.

> *"No problem can be solved from the same*
> *level of consciousness that created it."*
> *- Albert Einstein*

Here are a few potential corrective actions to consider:

a. Direct a manageably small area to be cleaned to each front-line supervisor's standard over the next two weeks. These areas will be inspected by the chain of command, up to and including the critique leader. The supervisor will be there to present his/her space and ask any questions that the group may have. When/if he asks, "What's the standard? What does he mean by clean? Should we scrub the floor with bleach? Repaint the walls?" the answer should be: "This is the working environment of the group that you directly supervise. Show us what **your** standard is."

b. Identify which of the areas referenced above meet your standards and then show all of the frontline supervisors these areas and explain to them that this level of cleanliness is the standard. Take photos of the area(s) for future use or to post for all to see. This is the standard and it has to be communicated.

c. How about inviting someone from the outside to tour your facility and give you feedback. Who? A representative from the the company's Technical and Performance Team, another region's vice-president, the plant manager of the cleanest plant in the company, or the even the company CEO.

We are not suggesting that these are the correct corrective actions. However, we do want to demonstrate the difference between accepting a corrective action that "sounds" acceptable and taking corrective action with some teeth.

If you were to implement corrective actions similar to these, you would receive feedback from your team that the corrective actions will interfere with the company's ability to perform its primary function of manufacturing Delorean Time Travel Machines. This will ALWAYS be the response. We don't have the manning, we don't have the resources, we don't have the time, we don't have the money. An entire book would be needed to address this common objection. For now, we will say that the solution is Leadership – leadership finds a way.

6. Corrective actions without a formal training program are like new recipes without a stove.

Training corrective actions and the pursuit of knowledge described in Chapter Two work together. Many of your corrective actions will include training. For example, training the front-line supervisors about the organization's standards of cleanliness. The critique may have revealed a handful of items that front-line supervisors must be (re)trained on, and this may require modifying the short-term training schedule.

Using the training program to support corrective actions from the critique process is largely self-evident, here are some elements of this interaction that may not be:

Check and double check your training attendance records. The corrective action does not direct "training for the flux capacitor operators who are able to attend Tuesday afternoon's training session." Track the absentees and schedule a makeup session. Even if there is only one person absent, do not scale down the training on the makeup session.

Update the required training for new hires frequently. For example, you may, after this critique, add a required signature from the Flux Capacitor Supervisor to a new hire's check-in sheet "Discuss the lessons learned from Critique."

You might be surprised to find how common is the following conversation during a critique:

Critique Leader: Louis (Flux Capacitor Operator), why didn't you shutdown the flux capacitor when the power level exceeded 1.21 gigawatts?

Flux Capacitor Operator: Sir, the procedure I was using directs a shutdown only when power output power exceeds 1.5 gigawatts.

Critique Leader: (Collapsing his face into his palms): We critiqued this same problem last year. What happened?

FCO Supervisor: The shutdown procedure was modified six months ago. We updated the revision log. When we implemented the revision log, all operators were trained on how to access it and when to use it. Sir, I am sure of it. In fact, I gave the training myself.

Critique Leader: Let me see the training binder. (Rifles through the binder to find the corresponding training sheet. Reads the sheet thoroughly with squinted eyes and a furrowed brow. Flips the sheet over to review the attendance log. Pushes the sheet to other end of the table where it slides into the hands of the FCO Supervisor.) Notice anything odd?

FCO Supervisor: Louis didn't attend the training.

Critique Leader: Louis, do you remember missing a training session last November and the makeup session in early December?

Flux Capacitor Operator: Insert one or more maddening responses such as:

- I was on vacation in November and I was working the back shift in December.

- I don't remember.

- I was hired in February.

- I was a Megatron tech until March; I wasn't part of that training group.

Ouch!

It takes effort to execute corrective actions completely. The price to pay of not doing so is high.

Reflection

Consider what you are trying to accomplish through this process. There are two outcomes that you are generating:

1. Getting to the bottom of a mistake that endangered (or injured/killed) personnel, impacted organizational performance, damaged the organization's brand, or any number of events that didn't go as planned. The critique process is designed to get beneath the initial assessment, which is usually draped on top of deeper rooted problems of an unknown depth.

2. Instilling the honest assessment process into the hearts and minds of the organization. The critique process is the most formal representation of a culture that you must strive to instill in every member of your organization. The five pillars, upon which your success will be based, will remain sturdy only if the organization is committed to self-reflection in a truly honest, and sometimes brutal, manner. The manner in which the critique process is handled is a true reflection of how the organization's leadership truly feels about this culture.

Sometimes these two outputs are mutually exclusive. The second consequence requires that the critique process is implemented in a way that encourages supervisors to embrace the process. To do this, you may have to throttle your intensity so you don't inadvertently do more harm than good. We've seen former Naval Officers make this mistake. They enter an organization like a bulldog and shove the process using the nuclear submarine standard down the throat of the organization. Although initially this strategy may result in a more effective critique, eventually it results in resentment and fear about the critique process. This is the exact opposite of what you are trying to create.

However, sometimes the mistake is of such a large consequence that you aren't concerned about building an organizational culture. People died. A large customer was lost. An avoidable mistake resulted in the loss of an unspoken amount of revenue. In these cases, you want everything that the Navy Nuclear Submarine critique process has to offer and you want right now. In these cases, it might be better to bring in a third party that can run the critique for you, which will reduce the probability that internal relationships will suffer.

SUMMARY

The critique process is an art. Nobody picks up a violin for the first time and performs masterfully. Unfortunately, many people that are assigned by their organization to lead the critique process consider the work to be administrative in nature. The administrative portions of the critique process are as necessary as strings on a violin. The real value of the critique is extracted through protracted effort, thought, and practice. Anyone who disagrees has a fundamentally flawed thought process regarding the objective of the critique process. We see this from time to time with organizations that use a version of this process to extract and evaluate the frequency of mishaps. This statistical process is not useless, but it is fool's' gold if it is not accompanied by a process that forces the organization to be brutally honest with itself. That deep stare into the mirror is scary and uncomfortable. As it is should be. An indication of a robust and successful critique process is that every critique should expose something about your organization's operations and/or culture that leaves the critique leaders jaw on the ground. This a big of pill to swallow, but it is wholly consistent with our cumulative experience spanning four decades and hundreds of critiques.

"Excuse me, can you please tell me how to get to Carnegie Hall?"

"Sure - practice, practice, practice"

Use the following templates to help you organize and record a critique.

CRITIQUE TEMPLATE

Location of Event:
Date/Time of Event:
Brief Summary of Events:
Location / Date of Critique:
Critique Leader / Senior Person in Attendance:
Attendees / Title or Position (Highlight any notable absences):

Timeline

(P#)	Time	Event

Problem(s) and Root Cause

Problem	Root Cause

Standard Root Causes (modify for your organization's needs):

1. Inadequate Supervision

2. Inadequate Training

3. Inadequate Standards

4. Inadequate Policy / Procedure

5. Inadequate Communication

Corrective Actions

Problem #	Short Term C/A	Long Term C/A	Completed (Initials / Date)

CHAPTER 8

ROOT CAUSES AND CORRECTIVE ACTIONS

The critique process described in the previous chapter requires substantial time and resources. The result of the critique process will be one of two things: a dividend or a tax.

If the critique process is effective and an organization successfully determines why the event occurred, the appropriate corrective action can be achieved. The avoided cost associated with recurrence and improvement in performance is the organization's dividend.

Failure to identify or implement effectively the appropriate corrective actions results in a tax. This tax will be the cost of the wasted resources from attempting to prevent recurrence and the cost of recurrence. The magnitude of both the dividend and the tax is larger than a first-blush analysis would indicate. The critique process can be invaluable method for revealing aspects of an organization's culture that would otherwise go unnoticed by the organization's leadership. Attempts to quantify this benefit are futile. A robust program that allows leaders to get to the core of their organization's culture is a perennial annuity that cannot be quantified. The flip side is true as well. If the critique process results in a program that yields no benefit to the organization, the associated tax is amplified by the negative impact that executing such an ineffective program can have on an organization's moral.

The process is a powerful one, and there is no "net zero" result. Either the process is effective and the organization reaps the dividend benefits or the process is ineffective and the organization is stuck with a hefty tax—and, likely, a damaged culture.

This is why the Nuclear Submarine community formalizes and invests heavily in the critique process. Formalizing the program is the easy part. The investment comes in the form of training the organization's leaders to successfully conduct a critique, identify root cause and corrective actions, and effectively implement those corrective actions. The stakes are too high to allow the leaders to rely solely on the formalized process that could, in theory, simply be handed to them in pamphlet form. Unfortunately, simply following the critique steps outlined in the previous chapter does not guarantee success. In fact, unless resources are devoted to training and mentoring the management team that will be leading these critiques on how to identify the root cause and implement corrective action, failure is a foregone conclusion.

ROOT CAUSE IDENTIFICATION

The heart of the critique process is the root cause identification. If you don't uncover and remove the root with a weed you pull out from your garden, that weed will grow back. The same is true of cultural and operational deficiencies.

Getting to the root is essential. And although this is not a new or controversial statement, many organizations fail to effectively identify root causes. The worst and potentially most dangerous scenario is when the leaders of an organization convince themselves that they do effectively identify root causes and corrective actions, when, in fact, they are only grasping the surface vegetation. Almost every operations organization implements a root cause analysis program, but few spend the necessary human capital in evaluating the effectiveness of the program. The size of your performance assessment team or the money spent on root cause analysis consultants does not guarantee success. Further, we must all be vigilantly on guard to protect ourselves from our internal biases. Often, we see what we want to see. Therefore, if I want to believe that my existing root cause analysis program is effective, it is very easy to subconsciously cherry-pick data that supports my belief.

Effective root cause analysis is challenging because it requires brutal honesty, which usually necessitates sacrificing egos for the sake of improving an organization's effort to achieve operational excellence. There is no doubt that this process is more of an art than a science. Artisans never master their craft—they climb a mountain without a peak—but they climb with a passion that borders on obsession to reach a zenith that they know doesn't exist.

THE RELATIONSHIP BETWEEN PROBLEM STATEMENTS, ROOT CAUSES, AND CORRECTIVE ACTIONS

If corrective actions are weak, chances are high that the root cause was weak. If the root cause was weak, chances are high that the problem statement was weak. The three elements string together like jigsaw puzzle pieces. The root cause attempts to identify the heart of the matter of the problem statement, while the corrective action attacks the root cause like a weed killer. Therefore, effective root cause analysis requires a thoughtful problem statement. For example, assume that you were at fault in a car accident. Specifically, you ran a red light and collided with the right-of-way car. You could use the following logic:

Problem Statement: You collided with another car.

Root Cause: You ran a red light.

Corrective Action: Stop at red lights in the future.

It is likely that you are missing something. After all, you already know you should stop at red lights. Is implementing that corrective action going to pay a future dividend for you? You were already trying to stop at red lights. In a critique, we would explore why you ran the red light. Was it intentional? No. Were you distracted? Yes. Why? You were reading a text message at the time. With this information we may conclude:

Problem Statement: You ran a red light while you were texting and driving. (The fact that you hit another car is simply bad luck. The problem was that your distracted driving caused you to run a red light).

Root Cause: Inadequate driving standards

Corrective Action: Install an application in your phone that prevents you from being able to text while driving.

Which of these two corrective actions are likely to reduce the probability of future car accidents?

There is often no "right" root cause or a single root cause. Reasonable people can disagree upon what "the" correct root cause is for a given circumstance; as previously stated, root cause identification is an art not a science. However, effectively identifying a root cause or root causes will usually allow the corrective actions to write themselves. Recall, after immediate corrective actions to stabilize a situation are taken, the subsequent corrective actions are assigned to address the root cause not the mishap that was critiqued because in most cases mishap is only a symptom or outcome of the root cause. Identifying corrective actions is like placing the few remaining pieces in a jigsaw puzzle. Provided the rest of the puzzle is assembled correctly (root causes) the remaining pieces fit effortlessly. Unfortunately, identifying corrective actions is the easy part and, like most assignments, their effectiveness is derived by their execution not their conception. We will discuss how to effectively implement corrective actions later in this chapter.

PREREQUISITE PHILOSOPHY

Prior to submerging the ship and usually before even leaving the pier, the Nuclear Submarine crew begins a process known as "rigging the ship for dive." This process checks the position of approximately 200 valves and switches. These items are the ones that ensure the submarine's watertight integrity is at full strength prior to submerging the ship. The ship is divided into approximately eight sections and each valve or switch in each section is independently verified by two people: one experienced, enlisted crew member and one qualified officer. Each valve and switch is listed, along with its required position, and both people initial next to each individual item. Their initial represents a promise to the Commanding Officer, the submarine crew, and their families that each valve or switch is positioned correctly. The 16 individuals are selected carefully for this evolution as it is a zero-defect process. If 199 valves are positioned correctly and one is positioned incorrectly, the family of a sunk submarine will take little solace in the fact that the rig-for-dive process was done 99.5 percent correctly.

The process of submerging a 6,900-ton ship is not taken lightly. In addition to the rig-for-dive process, the submarine crew conducts post-submerged checks of the entire ship. This means that when the ship first submerges, it holds steady at a shallow depth and then the entire ship is inspected for leaks, abnormal noises, or any other unexpected indication. On one particular Nuclear Submarine, a small leak of seawater coming into the ship was identified in Engine Room Lower Level during the post-submerged checks. A valve in the shaft seal system (the system that allows the propulsion screw of the submarine to penetrate the pressure hull while preventing seawater from entering the ship) was found out of position. The valve was a drain valve that was directed to be shut by the rig-for-dive procedure but was found open instead.

This is a big deal. The ship had a defect in a zero-defect program—rig-for-dive. The mishap was critiqued, root causes were identified, and corrective actions were taken. Of course, the two individuals that had initialed that this drain valve was shut faced disciplinary actions.

This example serves as a preface for a prerequisite philosophy that you must possess to allow the remaining message of this chapter to be meaningful. Specifically, although it is tempting to focus your root cause and corrective actions at the individuals involved, you will rarely get at the root cause by doing so. In the previous example, the two individuals that made the error were both superstar performers—recall, the people selected to rig the ship for dive are chosen carefully. However, time and again, all of the *initial* evidence points to the simple answer— two guys who were trained and aware of their responsibilities neglected their duties.

Right here. Right now. This is the difference between your tax and your dividend.

Learn to set your brain to the opposite position: *We have an organizational problem that requires corrective action to prevent recurrence. This is not a problem that is isolated to these individuals.* Accept now, that everyone around you will be literally or figuratively rolling their eyes to the tune of *Why are we making a mountain out of a molehill? Discipline the two people involved and let's move on with life.* If the organizational component that led to this mishap was easy to identify, we wouldn't be writing this book and organizations around the globe would be enjoying world-class cultures with sustained Operational Excellence. Not in every case (and that's important to keep in mind), but, in most cases, if you dig deep enough and ask the right questions you will discover an organizational element that contributed to the mishap. In all likelihood, this element has been a deficiency, or at a minimum a required improvement, for years. These organizational elements are masters of disguise, which is why it is so important to take advantage of mistakes and exert the energy necessary to sniff them out.

It is true that there are times when a mishap is as simple as it appears: Someone made a mistake, and there was nothing more to it than that. However, consider a spectrum with one end representing a mindset that the mishap was caused by the organization's culture, processes, or leadership. The other end of the spectrum represents a mindset that the organization is perfect, and this person made a mistake that was his alone and no one else's. Most leaders set their initial mindset to the latter and move to the former only if there is evidence that forces them to do so. In many cases, the leader will subconsciously overlook available but subtle evidence to support this mindset. We are challenging you to flip the script.

Always start with an assumption that there is an organizational element that needs to be addressed and force yourself to be convinced otherwise. This philosophy sounds simple. It is. This philosophy sounds easy. It is not. As a leader, you are proud of your organization, so why would you intentionally look for holes in your organization's structure because someone made a mistake? Because the ability to do so is precisely what determines whether you will pay a critique tax or earn a critique dividend.

In our example, the critique leader was able and willing to dig deep to identify the root cause behind the mistake. Because of his willingness and his practice in doing so, the critique leader was able to ask the right questions during the critique and the following information was discovered:

> ➤ The first checker hadn't been paid in the last two pay periods because of an administrative error at PSD (Personnel Service Detachment—The Navy's Human Resources department), and he had no savings to rely on. As a result, he had been unable to pay his family's bills this month.

> ➤ The second checker was the most-junior officer on the ship and was prone to seasickness. The weather was very rough that day. When the submarine is on the surface prior to submerging, the space he was assigned, Engine Room Lower Level, is the worst place to be for a seasick seaman. As a result, his ability to concentrate was substantially reduced.

> ➤ Although there were multiple ways to determine the position of this valve (physically checking, rising stem, and valve position indicator), the valve position indicator was swapped (shut meant open and vice versa). Several mechanics, including the one that found the valve out of position, knew of this material deficiency but had not reported it.

> ➤ The first checker was always assigned to rig Engine Room <u>Upper</u> Level for dive. However, the person who normally first checks Engine Room Lower Level for dive finished his tour of duty so the rotation was modified to compensate for this. Therefore, the first checker was not as familiar with this space for rig-for-dive.

> ➤ Both the first and second checker were assigned to the section that would be taking the watch when the ship submerged. Therefore, in order to complete their rig-for-dive and eat before taking the watch, they had to complete their checks in half the time they usually take or risk relieving their watches late.

> ➤ The first checker was part of the in-port duty section the night before. He stood the midwatch, and, although the midwatch is supposed to sleep from 6:00 p.m. to midnight, the Duty Chief had assigned him work during this period. As a result, he hadn't slept in over 24 hours while he was rigging Engine Room Lower Level for dive.

None of these factors relieve the individuals involved of their responsibilities in this mishap. Their initial is a promise, which they broke, and it is appropriate that they are disciplined and/or counseled accordingly. However, it is clear that there are several organizational elements that require attention. The good news is that once these items are addressed, the ship's processes will improve in matters that will impact more than future rig-for-dive evolutions.

ROOT CAUSE IDENTIFICATION

There are as many root cause identification methodologies as there are people with an opinion on the topic. In the previous chapter, we expressed our preference that almost every problem can be assigned one of five root causes. Yes, this is a very simplistic approach compared to many complex systems available for sale in the market. But, we like simple. It works. Here are the five:

1. Inadequate supervision

2. Inadequate training

3. Inadequate standards

4. Inadequate policy / procedure

5. Inadequate communication

However, we recognize that every industry and organization has unique attributes so you may decide to adopt a commercial program or modify our simple approach. That's your call, but here are some guidelines on criteria for determining root causes.

Effective Root Causes Lead to Actionable Corrective Actions

Recall that, ultimately, the root cause is the weed that must be eradicated. Therefore, a root cause that does not lend itself to corrective action is a weak root cause. Consider the example previously discussed about the rig-for-dive mishap. One of the contributing problems was that the second checker was seasick during the evolution. Although it may be intuitive to say the root cause here is: "Ensign Johnson is prone to seasickness," there are two problems with this root cause. The first is that there is no corrective action for seasickness (despite what any over-the-counter pharmaceutical company may have you believe). Once you are seasick, it's game, set, match. If I were to assign that as my root cause, one could argue that my corrective action could be "Do not assign Ensign Johnson rig-for-dive in Engine Room Lower Level." However, be on guard against corrective action that starts with "do not" or "will not." Even though the corrective action may seem logical as it is addressing the root cause of "Ensign Johnson is prone to seasickness," we've actually just assigned a corrective "inaction." Corrective inactions are not effective. We are looking for ways to attack an existing problem and that is best done with action. If your corrective action is a corrective inaction, revisit your root cause selection.

Effective Root Causes Prevent Recurrence Throughout the Organization

There are exceptions to every rule. Assume the "Do not assign Ensign Johnson rig-for-dive in Engine Room Lower Level" was evaluated as an acceptable exception to the rule of avoiding "do not's" as corrective actions. This corrective action is flawed for another reason as well. It is too short-sighted and narrow in scope. Let's assume Ensign Johnson breaks his leg next week and is no longer assigned to our submarine. Is the problem solved? Yes, and no. Yes, because Ensign Johnson will definitely not be performing any rig-for-dive evolutions. And no, because we could easily have the same problem with someone else, and we've taken no action to prevent this. This happened because our root cause didn't adequately address the real problem. The problem was not that Ensign Johnson was seasick. The problem was that we had a sick person performing a vitally important assignment. You see the difference?

Going forward we don't want anyone who is sick, from the sea or otherwise, conducting rig-for-dive. In this case, the manner in which we express the problem matters. Details matter. If we adjust the problem statement to: "An Officer performed an assignment critical to the ship's safety while suffering from a sickness that impacted his ability to do so without the chain of command's knowledge," now we have something that we can work with. We can tackle the problem from a process standpoint. The Ship's Diving Officer is the officer who assigns the rig-for-dive assignments. Since the evolution occurs on the surface, and many crew members get seasick when the ship is on the surface, would it be reasonable for the Ship's Diving Officer to screen rig-for-dive candidates based on whether they were susceptible to seasickness? That would be reasonable. We could tackle it from a communication perspective. The rig-for-dive evolution is a zero-defect evolution, and, therefore, if Ensign Johnson was seasick and trying to perform the evolution anyway, why didn't he tell anyone? Are we fostering a culture of excessive bravado where we put ego above reason? Maybe that is something we need address at training with the entire crew.

Effective Root Causes are Uncomfortable

Assume we concluded the following for the seasickness problem:

Problem: An officer on the ship performed a "safety of ship" duty with a sickness that impacted his ability to concentrate.

Root Cause: Inadequate supervision and standards. The ship's leadership has allowed a culture to develop that places bravado over reason. The ship's senior leaders have over-emphasized toughness and "getting the job done," and, as an unintended consequence, ship's personnel are reluctant to report when their health is impacting their performance.

Ouch. How difficult would it be to admit this? Especially, when the only initial data point was two guys mistaking an open valve for a shut valve. Root cause analysis can be performed two ways: the feel good way or the ouch-that-hurts way. The former results in a critique tax and latter pays a critique dividend. Creating a culture where the ouch-that-hurts becomes accepted and not fought will lead to an organization that is not afraid to face its problems head on.

THE WALLET THAT TOM FORGOT

What's a more mundane evolution than leaving your house? Grab your phone, your wallet or purse, and car keys. Make a final check that your socks match and you're out the door. However, today, Tom grabs for his wallet to pay for lunch and the dreaded empty feel of his back pocket places a pit in his stomach. He checks the other back pocket, his front pocket, his jacket pockets. Then he remembers that last night he made a late-night ice cream run, and he realizes that he probably left his wallet in those jeans. Tom spends the rest of the afternoon washing dishes to pay off his meal debt. As a result, he missed two critical company meetings that afternoon. When he returns home, Tom finds, as he suspected, his wallet in the jeans he threw on last night for an impromptu ice cream run.

Because Tom is a fictitious character in a fictitious and illustrative anecdote, his company decides to critique this mishap to reduce the possibility that he will ever again forget his wallet and be absent for critical company events. Tom's company follows the guidance in Chapter 7 and performs a critique. The root cause that was identified for the problem of "left home without wallet" was "inadequate attention to detail." The company assigns short- and long-term corrective actions. The short-term corrective action is that Tom's supervisor counsels Tom: a written reminder that his performance was unsatisfactory with a warning that continued instances of lack of attention to detail could result in further disciplinary action. The long-term corrective action is for Tom to "pay more attention to detail in the future."

A month later, the same thing happens. Damn. What happen?!

ACTIONABLE CORRECTIVE ACTIONS

Although counseling can often be an effective short-term corrective action to communicate and document substandard performance, these actions can rarely, if ever, be depended upon to fix or eliminate a problem's recurrence. Long-term corrective actions that come in the form: "(Name) will be more _____ in the future" are more common than you might believe and as ineffective as you might predict.

Why? For two reasons. The first is that Tom was trying to be attentive to the details of his morning already. The corrective action identifies the implementation of a process that was already in place, and we've observed that process to be fallible. The second is that long-term corrective actions are most often effective when they can be implemented by anyone in the organization. Telling Tom to be more attentive to details in the future does not address how to prevent this from happening to other people in the company.

"Tom will be more attentive to detail in the future" is code for one of three things:

1. We have not identified an actionable root cause.

2. We don't want to change anything fundamental about the way we do business.

3. Tom has proven to be a screw-up (but not bad enough to terminate), and we are not going to change our processes because one guy can't get it together.

None of these reasons are particularly strong. The first two are weak and counterproductive and third is not uncommon and is understandable—but not particularly effective. If it was effective, Tom would be performing at a higher standard by now.

In fact, "inattention to detail" was banned for a period of time by the Nuclear Submarine community as a root cause for these reasons. The Nuclear Submarine community's senior leadership were encouraging Commanding Officer's to dig deeper and "inattention to detail" is rarely a deep enough root cause to prevent recurrence.

LEVEL OF MAGNITUDE

Now may be a good time to discuss priorities. Every organization will have mishaps. However, certain ones are unacceptable at almost all costs and others are acceptable at certain intervals beyond a predetermined cost. For example, how much effort (time and resources) is an organization willing to devote to ensure that none of its employees forget their wallet, keys, phone, etc. at home. Likely—very little. But IF the company leaders did designate this type of mishap as one that is absolutely unacceptable, what do they do? They've warned Tom, they trained Tom, and he is still getting it wrong a time or two a year. Then Steve from marketing makes the same mistake. What is going on?

The corrective actions, derived from the root causes, were applied to a small section of the organization—in this case, only Tom. This is natural but not effective. Why do organizations make this mistake? Why don't they cast the net of corrective actions to a larger section of the organization—if not the entire organization. The answer is simple (but incorrect): because the other parts of the organization did not have a problem. If it's not broke, don't fix it. Right? This brings us back to root cause analysis. Let's assume that the problem of forgetting one's wallet, keys, or phone is absolutely unacceptable. We must ask ourselves why Tom made this mistake—the root cause. Our first two efforts at root cause identification — "inattention to detail" and "lack of training"—seem to have missed the mark. We discussed why "inattention to detail" was a poor root cause, but what about training? In our example case, the corrective action that addressed "lack of training" missed the mark for two reasons.

The first was that it didn't apply the training to the entire organization. If we assessed that Tom was making this mistake because he was not properly trained on how to leave his home, then would it not be reasonable to assume that others would share the same training deficiency?

The second reason is that if the training cannot be validated by testing or measuring post-training performance results, it will be inadequate to correct the problem.

This rabbit hole has no end. We could train everyone in the department, just like we trained Tom, but, clearly, that would not be enough. Tom had the training, yet he made the mistake again six months later. We could then begin to ask ourselves: *How big is the problem?* Tom and Steve's mishaps have risen to our level of attention because they resulted in a business problem—Tom missed a critical meeting because he was washing dishes, and Steve missed critical phone calls. However, how many people are forgetting the phone, wallet, or keys, but we don't even know about it because nothing untoward occurred as a result. If the mistake has been identified as one that is unacceptable at any costs, maybe we should be checking people on a daily basis. But then who checks the checkers?

We hope that you can see how easy it is to start chasing your tail when trying to solve a problem with an unclear or incorrect root cause.

We haven't yet discussed procedures or supervision. It is tempting to focus on correcting the actions of the individual that made the mistake—fix that person, counsel that person, train that person. The corollary to this tendency is to consider training—either to an individual or a group—as the fallback solution. This is rarely a root cause fix.

NUCLEAR SUBMARINE ROOT CAUSES

The list of five root causes that we referenced earlier in the chapter are NOT standardized root causes used by the Nuclear Submarine community. The root-cause analysis thought process is deeply ingrained similarly in the minds of every submariner, but the application will vary slightly as a function of each Commanding Officer's philosophy of root cause analysis. We present these root causes as representative of those used in the Nuclear Submarine community, and those that are preferred by the authors. Let's take a look at each one.

Inadequate Supervision

Supervision is an essential part of obtaining Operational Excellence in any organization. However, the effective application of supervision is both subtle and nuanced. In every aspect of an organization's operations, the element of supervision must be handled carefully, thoughtfully, and deliberately. Often we don't provide our people with enough supervision; sometimes we provide them with too much. Who is supervising? What type of supervision are they providing? Where are they physically located while supervising? Are our supervisors trained to identify unusual circumstances?

Anyone who has ever served aboard a Nuclear Submarine is uniquely familiar with the term "field day." It sounds fun, but, unless you enjoy cleaning and painting, it's not a recreational event. "Field day" is a Navy expression for a time set aside for cleaning and preserving the submarine. No one is exempt from field day. The entire crew sets aside its work and participates. On a particular Nuclear Submarine, the Chief Mechanic instructed his two most-junior mechanics to paint a section of Auxiliary Seawater Bay—the bowels of the ship.

His instructions were direct. "Petty Officer Johnson and Smith. You are going to spend the next four hours painting Auxiliary Seawater Bay. Let me show you exactly where."

As the most-junior mechanics on the ship, Petty Officer Johnson and Smith, knew very little about the ship's operations. The Chief Mechanic knew this, but he also knew that their able bodies were well equipped to paint. The Chief showed them the areas that required painting, provided them the paint they needed and the tools (brushes, drop cloths, rags, etc.).

"Do you have any questions for me?" the Chief asked.

Petty Officer Johnson answered for them both. "No Chief. We are on it."

The Chief left the area and spent the remaining four hours in Engine Room Upper Level shooting the breeze with the Electrician Chief while periodically barking out orders to the cleaning crews in the immediate vicinity.

Four hours later, the Chief returned to Auxiliary Seawater Bay to inspect the painting job that he anticipated would be completed. What he saw wasn't good. Although Petty Officer Johnson and Smith had clearly worked hard, they also made a mess that at first glance would be unrecoverable. They painted EVERYTHING—pumps, motors, motor controllers, damage control panels, the installed phone system.

There are certain components in the Engine Room that you don't paint or at least not with the Seafoam Green color that these two mechanics used. The Chief had intended for them to paint the bulkheads, the bilges, and the piping. The Petty Officers were confused and frustrated when the Chief lit into them. "How stupid are you guys? Do you really think that I wanted you to paint over the red flooding alarm or the pump motor controllers? Are any of these items throughout the boat painted green?"

Maybe these mechanics should have known better. Maybe these mechanics need formal training on how to preserve different areas of the boat. Maybe these mechanics deserve to be punished. Maybe they deserve to lose their weekend and work on undoing their mess.

However, the problem that "junior mechanics inappropriately painted many items in Auxiliary Seawater Bay" has only one root cause worth addressing: inadequate supervision. Sometimes the most important decision a leader has to make is deciding where to physically be. The Chief Mechanic chose the comforts of Engine Room Upper Level but had he chosen to watch, or at least periodically check on, his junior mechanics in Auxiliary Seawater Bay, this mishap would have been averted.

Inadequate Training

There is a flurry of activity on the day prior to a Nuclear Submarine getting underway. The preparations are immense and require all hands to participate. One of the assignments is to prepare the "bridge bag." The bridge bag is an oversized gym bag that is full of the tools that the Officer of the Deck needs when he stations himself on the bridge. The bridge is the area atop of the Nuclear Submarine sail where the ship is controlled from during surface operations. Each component in the bridge bag is critical. To ensure the bridge bag is prepared fully, there is an inventory checklist.

On this particular submarine, the Navigation Electronics Technician assigned bridge bag preparation was Petty Officer Jones. Petty Officer Jones had been a member of the ship's crew for almost two years and had prepared the bridge bag several times before.

One of the items required is a handheld bridge-to-bridge radio. This radio allows the Officer of the Deck to listen and talk to other ships. His ability to do so is imperative to the ship's ability to navigate on the surface. The checklist stated:

Bridge-to-Bridge Radio (1). Ensure battery is fully charged and radio is set to scan channels 16, 18, 43, 65 [Note: these are fictitious settings.] Test radio to ensure transmit and receive functions work properly.

As he set the channels, Petty Officer Jones accidently set the radio to scan channels **17**, 18, 43, and 65. It wasn't intentional; he simply entered 17 instead of 16.

Fast forward to the ship's surface transit. About 15 minutes into the underway, the Officer of the Deck noticed a large tanker that was travelling on a collision course with the submarine. The geometry was such that the submarine had the right of way and the tanker was required to slow down or turn away. If neither ship maneuvered, a collision would occur in 15 minutes. Fifteen minutes is but a few seconds when it comes to surface operations. The Captain of the tanker ship was trying to communicate with submarine on Channel 16, but the bridge-to-bridge radio wasn't scanning Channel 16. The Officer of the Deck eventually realized that the radio was programmed improperly. Seconds were turning into minutes while the Officer of the Deck fiddled with the radio. He had grown so accustomed to the radio being programmed correctly that he could not figure out (under pressure) how to transmit/receive on Channel 16. Ultimately, it is only a few button presses that the average civilian could figure out, but with pressure mounting and his experience low, the Officer of the Deck called for a Navigation Electronics Technician to report to the bridge to fix the radio. To avoid a collision while he scrambled to fix the radio, the Officer of the Deck ordered the submarine to a slower speed. Of course, so did the Tanker. As a result, they remained on a collision course. Turning right or left was not an option because the water was too shallow outside of the channel for the tanker or the submarine. The Electronics Technician was on the bridge moments later and programmed the radio correctly. The Officer of the Deck and the Captain of the Tanker were then able to communicate and coordinate their ships' speeds to avoid a collision, but the resulting CPA (closest point of approach) was under 100 yards, essentially one submarine length. This near miss needed to be critiqued.

During the critique there was plenty of "blame" to go around.

➢ The backup bridge-to-bridge radio in control was broken but not reported as such through the chain of command.

➢ The contact coordinator was newly qualified and was tentative about his reports of the tanker. He had information that should been reported much earlier, which would have allowed the Officer of the Deck more time to deal with the situation.

> ➤ Petty Officer Jones made a mistake when programming the radio while preparing the bridge bag.

> ➤ The Officer of the Deck was not proficient in operating the bridge-to-bridge radio.

All of these problems were addressed during the critique; however, the Commanding Officer supported a root cause of "insufficient training." Anyone could have made an error in programming a radio because we are all fallible. New contact coordinators are timid. Material deficiencies happen. Ultimately, the following was determined:

Problem: The ship had an avoidable and unacceptably close CPA with a tanker due to the ships' inability to talk to one another.

Root Cause: Insufficient training. The Officer of the Deck was not proficient at operating one of the the most important instruments on the bridge.

Corrective Action: The Navigator will train the wardroom (all of the ship's officers) on the use of the bridge-to-bridge radio. Prior to taking the watch as Surfaced Officer of the Deck, the officer would have to display to the Commanding Officer the ability to use all of the radio's features.

The Commanding Officer's objective was to never have a near collision because of the Officer of the Deck's inability to use the full functionality of the bridge-to-bridge radio. Wardroom training, followed by a validation of that training, was the corrective action that was used to attack the problem and root cause. Note that the training was provided to all of the officers, not just the officer involved in the incident. Further, demonstrating proficiency in operating the bridge-to-bridge radio to the Commanding Officer was added to the Surface Officer of the Deck's qualification card to ensure that this problem doesn't recur five years from now. This is a good example of addressing training deficiencies.

Inadequate Standards

There are three enlisted watchstanders and one officer supervisor (Engineering Officer of the Watch/ EOOW) in the Maneuvering Area (the controlling station of the Nuclear Propulsion Plant). The three stations in the maneuvering area are the steam plant, the reactor plant, and the electric plant. The watch is 8 hours long, followed by 16 hours off, followed by 8 hours on, etc., for months on end. The watchteam is often required to train each watch on a particular topic, and the conversations that occur in maneuvering cover every topic you can imagine . . . and then some. However, from the very beginning stages of the training pipeline, watchstanders are taught to always be scanning their panel. You can talk to the EOOW, but there is no need to turn around and look at him to do so. Keep your eyes on your panel. This is more difficult to do than it sounds but it is emphasized to a high degree during the training pipeline. Unfortunately, it is human nature to drift from standards if there is not force applied to prevent them from doing so.

On this Nuclear Submarine, the Electrical Operators had grown accustomed to turning their chair at a 45-degree angle away from their panel so they could talk and see the rest of the members of the watch station. While having a conversation with the EOOW about the amount of air available on a submarine if we stopped making oxygen, the Number Two Turbine Generator (one of the turbines providing electrical power) tripped off-line. Because he was looking at the EOOW and not his panel, he did not notice the trip. Alarms began to sound as the second-order consequence of the trip resonated throughout the power plant. Had the Electrical Operator been watching his panel there were indications that would have allowed him to act and prevent the Number Two Turbine Generator from tripping.

Maybe this appears like a slam dunk. Inadequate supervision. Why didn't the EOOW enforce the standards that had been ingrained into all of the crew from day one? What about inadequate standards of watchstanding? The electrical operator had been in the Navy for 10 years. He knew the standard and choose to ignore it. Discipline the EOOW and the Electrical Operator and then get on with life. Right?

The Commanding Officer chose to dig deeper and force himself to be as brutally honest as he could. Recall the prerequisite philosophy that we discussed: Set your mind to a starting point that there is an organizational element behind each mishap until the evidence overwhelmingly proves you wrong.

The Commanding Officer asked the Executive Officer, "XO, when was the last time that you were in maneuvering?"

The XO replied, "Yesterday afternoon. I went to maneuvering to remind the group that there was an all-hands training on Crew's Mess immediately after their watch."

"How was the Electrical Operator seated at that time? Was he looking at his panel?"

The XO grimaced, "Actually, he was seated at about a 45-degree angle away from his panel. Come to think of it, he asked me a few questions about the training and was turned completely away from his panel."

The Commanding Officer then asked the Engineer Officer, "Eng, when was the last time you reviewed a casualty with the maneuvering area watchstanders?"

The Engineer Officer knew where this was going. "Last night. As I recollect, the Reactor Operator turned towards me every time he said something. The more I think about it, I see what you driving at; leadership hasn't been enforcing the standards of watchstanders facing their panels."

The Commanding Officer was glad that the Engineer Officer picked up on where he was going, because he didn't want to have to share his experience on his last trip to maneuvering when the watchstanders were turning away from their panels to talk to him. *If I am not enforcing the standard, the watchstanders will logically take that as implicit permission to deviate from the standard.*

The Commanding Officer responded to the Engineer Officer, "Eng, you are exactly right. This problem falls squarely onto the laps of the ship's senior leadership. If we all have observed deviations from this standard and none of us have corrected it, this event should come as no surprise. We allowed this to happen. No, we made this happen."

Leaders have to reinforce the standards at all times and recognize that by not enforcing the standards, they are setting new standards. Because this Commanding Officer recognized that he and his leadership team were the root cause (rather than just blaming the watchstanders), they were able to correct a poor practice and improve the Operational Excellence of their ship.

Inadequate Policy / Procedure

Sometimes everyone does the job exactly as it should be done, until something untoward occurs anyway. For example, the Engine Room watchteam is starting up the turbine generator lube oil system. They perform as the consummate professionals they are. There is a mechanic reading the procedure and a mechanic performing the procedures. Several moments later, there is lube oil spraying from a vent line. The mechanic quickly shuts the vent valve, but now they have quite a mess on their hands.

During the critique, the senior leadership identify three things:

1. The initial conditions of the procedure direct the system to be lined up in accordance with Figure 2-1. Figure 2-1 is a valve line-up that lists every valve and switch in the system and its expected position (open, closed, on, off, etc.). The vent valve is listed as "OPEN."

2. The procedure never directs shutting the vent valve.

3. The procedure was taken directly from a new revision that had recently arrived on board from Naval authorities and had not previously been used.

Even with his mind set in the "prerequisite philosophy", the Commanding Officer had to agree that the significant root cause was procedural. However, as is often the case, there was another root cause aside from the obvious one. Specifically, the ship's process for implementing new procedures needed improvement. This was a perfect example of why each revision needs to be read, trained on, and thought through before it is implemented. Even procedures provided by naval technical groups can have errors. Therefore, the ship's procedure for implementing new procedures required improvement as well.

Inadequate Communication

There is probably not a mishap under the sun that doesn't somehow involve a communication gaff that contributed to the mishap. Usually, miscommunication contributes to a mishap, but sometimes it is the root cause. We want to be careful about using inadequate communication as a fundamental root cause because it can be difficult to address. Also, if there is a miscommunication problem, or a trend of miscommunication problems, there is usually a more fundamental problem to uncover.

The Engineering Department routinely runs casualty drills to ensure the department is ready to handle any situation that may befall it. It takes a lot of manpower and planning to run casualty drills. There is a drill team that "runs" the drills—they provide simulations, monitor the watchteam's performance, and intervene, if necessary, to prevent any undesired plant effects. The Engineer and the Engineering Department Master Chief present a drill package to the Commanding Officer at least one day in advance. Some casualties will limit propulsion; others will require the ship to start at a deep depth; and others require the ship to be at periscope depth. The point is that engineering drills impact more than just the 55 men in the department—they impact the entire submarine.

When a submarine is submerged it is assigned its waterspace. Think of a box that moves in the ocean: the box can change sizes, it can be deep or shallow, and it can travel fast or slow or be stationary. It is the Commanding Officer's responsibility to keep his submarine in that box. This allows the submarine community to put more than one submarine at sea at a time and eliminate a collision risk. It is analogous to Air Traffic Control around an airport. Each submarine at sea has its own box that it must stay in. If the submarine is transiting, the box continues to move, even if the submarine doesn't. Therefore, if you are going to run engineering drills that will limit the ship's propulsion or even require the ship to stop, the submarine would be wise to start that drill period with the submarine towards to the very front of the moving box. This will result in the submarine slowly drifting towards the back of the box during the drill period. When the drill period is over the submarine can then race back towards the middle of the box where it typically spends its time.

One morning aboard a Nuclear Submarine, the Engineer has his drill team assembled at 0500. The Engineer heads up to the Commanding Officer's stateroom to invite him to the drill brief. The Commanding Officer is wearing a scowl that could frighten a honey badger into submission.

"You are not running any drills today. Go talk to the Navigator and figure it out."

Boom! The Commanding Officer slams his stateroom door shut. The Engineer stands on the other side of the door in disbelief.

The Engineer goes into the control room to talk to the Navigator. Before he says a word to the Navigator, the Engineer identifies the problem. The ship is operating in the back of the box.

What the heck! They had all night to work their way to front of the box.

By the looks of the Navigator's face, the Commanding Officer had already gotten to him.

The Engineer holds his hands out in front of him and sighs,

"What happened?"

The Navigator launches into a list of items that they had to complete on the midwatch, all of which required the ship to operate slowly and/or at periscope depth. The list was mostly requirements of his department and the Weapons Officer.

"Why would we do all of that the night before an engineering drill set?"

The Navigator responds, "I didn't know that you needed the ship to be ahead of PIM to run today's drills." ("Ahead of PIM" is submarine speak for in the front of the box.)

The Engineer is stunned.

"The Weapons Officer gave me a list of items that he said needed to be done. I assumed that you two had spoken."

We'll stop the example here. I think you get the picture.

The three department heads aboard this Nuclear Submarine were clearly not communicating with one another. Three men's failure to communicate and hundreds of man-hours are impacted. The crew's morale is impacted, and, of course, no engineering drills were run that day. Why? What was the root cause of the cancellation of engineering drills? Inadequate communication.

> *"What we have here is failure to communicate."*
> - Prison Warden from the 1967 film, Cool Hand Luke

OWNING THE PROCESS

Root cause analysis and corrective actions are the most challenging topics in this book to address for several reasons.

The first is that the premise is so well understood by leaders. This may lead you to believe that it is a good thing, and it is – ish. There isn't a management team on the planet that doesn't believe in root cause analysis. Unfortunately, understanding it and getting it right are two separate matters. In fact, some leaders think that root cause analysis is so important that they hire a full team to tackle issues as they arise. This decision essentially "outsources" the root cause analysis process—or, at best, it delegates the process away from senior management. We can't and won't say that this process is ineffective. We can say that we have not yet seen the process be effective. If there was a team that parachuted onboard the submarine every time we had a critique-able event and handled the root cause analysis, we can assure you that the result would have been a critique tax. When the senior leaders are directly involved, the process works better. It just comes down to ownership. As a leader, you have to own the operations, and you have to own the root cause and corrective action process.

Secondly, the tax and dividend differential is very challenging to quantify. A critique that lasts two hours and spits out ten corrective actions to address two root causes will, without exception, sound very strong.

We've identified that two individuals failed to perform their jobs to a satisfactory level because they are not committed to the standards of excellence that is this company's benchmark. As a result, the employment of both personnel has been terminated, effective immediately.

This certainly sounds like it was written by someone that cares about the company and about upholding its standards. Unfortunately, a critique that doesn't get to the real root cause does not read or sound much different than one that does.

Lastly, most leaders think they are the ones that "get" root cause analysis. This is similar to the phenomenon that most people believe they possess above average intelligence. You can't help people who don't know they need help. On the other hand, not everyone does need help. However, it is clear to us that most companies do not successfully implement root cause analysis. As a result, it is the employees who pay the price. Their working environment has a weaker culture than it could. Their company has a weaker performance output than it could. They continue to work in an environment short of what could be considered Operational Excellence.

CHAPTER 9

PENDULUM LEADERSHIP

Nothing that we discussed in the previous eight chapters is possible without strong leadership. We've discussed the elements of the Nuclear Submarine's culture of operational excellence. Culture is the product of how members of an organization interact with each other. There are countless iterations of interactions within even the smallest of organizations but none more critical to culture strength and operational excellence than the communications initiated by the leaders.

What makes a strong leader? This is the question that is discussed in classrooms around the world. It is the essay question that we all have had to answer at least once in our lifetimes. It is also an unanswerable question. Asking "What makes a strong leader" is like asking "What makes a good light." The answer is entirely contextual. In this chapter we are going to take a look at the leadership required, in three scenarios, to create, improve, or maintain the elements of operational excellence that we have discussed in this book.

These three scenarios represent a spectrum of the levels of operational excellence into which you may find yourself entering. They are:

> ➢ The organization that is falling apart and is in danger of dissolving—the broken ship

> ➤ The organization that is doing "okay" but has grown content with being just okay—the mediocre team

> ➤ The organization that has a tremendously positive reputation—the superstar team

The Perennial Pendulum

Given the accelerated ebb and flow of personnel changes aboard Nuclear Submarines, it is commonly accepted that individual submarine performance tends to follow a pendulum pattern. The worst ships are getting better, and the best ships will struggle to stay on top. It is important to note that in the context of Nuclear Submarine performance and culture, "poor performing" boats are described as such only because the standard of performance in the community is relative to the other submarines' performances. These comparisons result in a necessary but unspeakably high standard.

Leadership is integral at all levels of the chain of command. On a Nuclear Submarine, the culture is determined by and large by the Commanding Officer, but not all leadership is driven from the senior person in an organization. In fact, we have intimate experience with a Nuclear Submarine crew that was able to achieve every accomplishment available, in spite of— not because of—the Commanding Officer's leadership style. In this scenario, the ship just happened to have a group of Department Heads who were without peer in the fleet. This example is provided as an optimistic perspective, not a disparaging one. Sometimes, the strength of an organization can come from an inner circle of leaders that can compensate for leadership shortfalls at the top. Other times, the cultural change comes from the bottom up. Where there is darkness, where there is uncertainty, and where there are cultural deficiencies, there is an opportunity for leadership at all levels. Change can and often does come from just a few motivated individuals who are committed to driving the organization to operational excellence. Every scenario is unique and has multiple paths to success. These scenarios are presented not as full case studies, but rather as summaries of what entering an organization at different phases can look like.

THE BROKEN SHIP

A Nuclear Submarine was involved in a collision at sea. The damage to the submarine and the tanker that it collided with was substantial. The investigation revealed that the most senior leaders on the ship had not maintained a culture based on the principles we have described in this book and that allowed a failure of the watchteam to keep the ship safe. The ship's culture was not a total disaster, but there was enough erosion of the high reliability principles that a fairly routine surface transit turned into a collision between a nuclear submarine and a foreign tanker. After the dust had settled, the senior leadership was relieved and replaced with interim leaders who were assigned to "fix" the ship's culture. The individuals assigned to the ship were hand-selected because of their track records of success. Once the new leaders were in place, there was a collective sigh of relief that things would now be better. Unfortunately, cultural change is not a light switch. Even with stronger leaders placed in position, the remnants of the past culture would not go quietly.

The Executive Officer sensed that the ship had cultural problems that were more deep-rooted than could be explained by the recent leadership combination. His gut feeling was that the ship needed more than just a few new leaders in place. There was work to be done to turn the culture of the ship around. The pendulum had not yet reversed course. During the ship's recovery, an electrician received a powerful electric shock due to poor work practices in the engine room. If there was any doubt that simply putting strong leaders in place would reverse the direction of the pendulum, this event removed that doubt. The ship's culture required damage control . . . and fast.

What is a new leader supposed to do to reverse the direction of the pendulum?

Although there is no one right answer, the new leadership team on this Nuclear Submarine began to focus on the basics. The basics of the basics—a strategy that was not unlike the "first principles" we discussed in Chapter Two. The leadership team worked to evaluate the existing policies and processes and re-place them, if necessary, with fundamentally strong ones. In doing so, they tackled each of the five principles directly or indirectly. For example, when it came to knowledge and learn-ing, the training and qualification process was turned on its head. Not with malice, but the help of the source documents (Engineering Department Manual), the Executive Officer re-viewed each training and qualification requirement—painfully, one by one, and then evaluated whether or not the training and qualification program met that standard. The program was sub-stantially more robust when that process was completed, but it wasn't more administratively burdensome. In fact, the oppo-site was true. The Executive Officer discovered that the ship's leaders were focused more on the administration of programs than they were on the implementation of these programs. In an attempt to make their programs "look good" to external audi-tors, they were choosing form over function.

There were substantial second- and third-order positive impacts resulting from the Executive Officer's review. Word spread quickly that the Leadership was reviewing all of the ship's programs. Further, the crew was learning that the Commanding Officer and Executive Officer were forcing all conversations about ship's programs back to the source document. As a result, program managers began to review their source documents for requirements to ensure that their programs met those standards to get ahead of the Executive Officer's review.

The Commanding Officer and Executive Officer spent time training the ship's Officers and Chiefs on how to monitor and repair basic watchstanding practices, such as observing people turn over their watch—What information do they share? How do they share it? etc. However, during these training sessions, the Executive Officer began to doubt whether or not the leaders of the Nuclear Submarine recognized that they had a problem. Did they honestly believe they had a cultural issue that required effort on their part to fix? If they didn't, was he wasting his time?

When bringing change to an organization with a weak culture and recent poor performance, it is worth taking the time to assess whether the members of the team truly believe they have a problem. Our experience is that many leaders in this situation do not believe the team's problems stem from their own personal performance or a culture problem. Instead, a victim mentality arises quickly and grows deep with thick roots. Of course, most of the ship's leaders would not express this victim mentality directly to the senior leadership. However, if they were to express their version of the ship's narrative, it would sound something like this:

"We have always been a strong performing boat. We had one untoward event that resulted in the collision, but that was a freak accident, and if there was anyone to blame, it was the former Officer of the Deck during the collision, and he was fired. The most recent event of an electrician receiving a shock was another freak accident that was totally isolated from the collision and was the result of a few people in the Engineering Department trying to take a shortcut. We are suffering through this painful wire brush scrub unnecessarily; we are ultimately a good boat that has suffered through some bad luck."

This narrative is very typical of poor performing teams following an incident, and this attitude can substantially block efforts to improve the culture. Organizations that suffer from poor performance rarely volunteer to hunker down, roll up their sleeves, and fix themselves. Instead, they will often generate excuses and blame the larger organization's leadership and policies that have resulted in their poor performance.

The Executive Officer in this case eventually felt it necessary to address this gap in narratives between the senior leadership and the ship's midgrade leadership. Bridging this gap would not be easy and is laden with traps. If the Executive Officer attempts to win over the crew, he may be tempted to "go native" and become a sympathizer with their victim mentality. Although, this may result in the front-line supervisors "liking" the Executive Officer more, the chances for real organizational change drop to near zero.

Falling into this temptation might sound something like, "Look, I think you guys are a strong crew that operates with high standards. Before the collision, you enjoyed a much-deserved strong reputation, and you achieved everything that was asked of you. The reality is that we now find ourselves under a microscope that isn't going away. The best way for us to handle this is to go through the motions that are expected of us. We'll improve our processes a bit, and soon enough, we will be back to running our submarine the way that we know how.

However, I need you guys to understand that, right or wrong, we are going to have to deal with some BS for awhile. If you make an honest effort to get through this BS with me, I will do my best to accelerate that process."

Some might argue that this approach isn't a horrible one. At the end of the day, the Executive Officer needs the Nuclear Submarine's leadership to participate in a program to improve the day-to-day level of performance. If he acknowledges that they have done nothing to deserve this, he may receive stronger participation in his programs; however, it is unlikely that actual operational change will occur.

An alternative approach is the following:

"There is no doubt that you are all professionals. However, as a team, we have some work to do. The day-to-day standards that we are executing are not satisfactory. However, let's forget that we are under the microscope, and other boats aren't. Other submarines don't matter to us right now. Let's make our biggest competitor ourselves. Instead of making excuses and feeling sorry for ourselves, let's pull together and use this opportunity to shore up the operational elements of the way we do business. Regardless of where we stand relative to other boats, let's commit to a program of continuous improvement. Let's use this opportunity to dig deep and find out more about ourselves and what we can improve on. No, it won't be comfortable, and it will require setting aside our egos, but let's use this opportunity to make this ship the best submarine in the fleet. Submarining is a tough business with a very high standard. I need each of you to ask yourself if you are willing to accept that our biggest competitor is ourselves. We are not competing against anyone else. "

This approach avoids the "you suck" message but doesn't cave to the victim mentality either.

Ultimately, the Leadership team focused on stepwise improvements in standards. The standards that were emphasized with extreme and constant pressure were procedural compliance, formality in communications, integrity, and an aggressive critique program. These were non-negotiable.

To accelerate the culture change, the Executive Officer also implemented changes in the ship's operations that would produce incremental change. These changes were subtler and less immediate, but analogous to pushing a large boulder. The objective wasn't to push it across the finish line tomorrow, but rather to overcome the inertia of the boulder and get some incremental forward progress that would, in time, result in substantial changes in the ship's culture.

Examples of these incremental efforts included:

- Adjusting the ship's qualification program by creating a select list of people who were authorized to sign qualification cards.
- Bringing energy and a more dynamic nature to the crew's training program by posting the results of each examination and providing incentives for high performances.
- Expanding the officer's training program to include leadership and management topics, which were augmented by a required reading list of leadership books. The content of these books was discussed in open forum sessions among the officers.

The specifics of the turnaround led by this Leadership team are deserving of a book unto itself. The takeaways of this snapshot are:

- Replacing leadership doesn't automatically result in cultural changes.
- Don't expect the incumbent leaders of a poor performing unit to acknowledge either the poor performance or their responsibility in that poor performance.
- You can't tackle everything at once. Select a few items that are non-negotiable and apply daily pressure to these elements with the objective being an immediate and stepwise improvement in these items.
- Implement policies that overcome existing inertia in other areas that may not result in immediate change but that will create daily incremental progress that will, in time, yield significant improvements.
- Challenge the group to accept their competition as themselves. "You may not be 'bad' but how 'great' can we become?"
- Acknowledge that it is possible for a group of strong performers to combine into a team that underperforms. It happens all of the time in sports. The current poor performance is not a personal indictment on anyone's abilities, but rather a challenge to better work together as a team.

THE MEDIOCRE TEAM

Commander "Jones" takes command of a Nuclear Submarine. The submarine has been a steady performing boat for years. It has received average scores on all of its inspections and has a reputation for being a mediocre boat—nothing horrible, nothing great. Expectations for the boat's performance tend to be low. The crew is considered reliable enough to stay out of trouble, but not talented enough to take on the tough missions and assignments.

Commander Jones finds himself in a tight spot. He possesses a Type-A personality and has enjoyed a Naval career that is teeming with awards and recognition. His goal is to raise the standards of the boat from "mediocre" to "excellent"—one of the toughest leadership challenges of all. The pendulum is steady at its equilibrium position without any noticeable movement in either direction. This situation is probably the most common of the three scenarios that we are presenting in this chapter—an organization that is getting by but underperforming from its potential.

Commanding Officer (CO) Jones, wisely spends the first month of command observing the ship's performance while resisting the temptation to make too many changes. Although his aspirations to transform this mediocre submarine into one that exemplifies operational excellence remain, he is cognizant of the "first do no harm" principle. Sometimes sweeping changes in an organization that is barely getting along can push a dangling pendulum in the wrong direction. His approach will have to be thoughtfully tactical.

The CO decides he first needs to motivate the ship's senior leaders. He gathers the Executive Officer, the four Department Heads (Engineer, Navigator, Weapons Officer, and the Supply Officer), the Chief of the Boat, and the Engineering Department Master Chief.

"Gentlemen, it's been a month since I took command, and I want to spend some time communicating my observations to the ship's senior leadership. So here we are.

"Let me start by saying that I admire the professionalism of each and every one of you. The leadership—you—aboard this ship has performed exemplary. The submarine runs well on a day-to-day basis. However, I think we can do better.

"Nav, do you know why I think we can do better?"

This question catches the Navigator off guard. He stumbles through a response, "Our formality is a bit weak, and our planning process could be improved? I'm sure there are other areas, but those are the two that come to my mind."

The Commanding Officer replies, "I'm not going to disagree with that."

"Eng, what about you? Why do you I think I believe we can be better?"

"I'm not sure, Captain. We are working as hard as we can and always meet the requirements placed on us."

The Commanding Officer pauses for a few seconds. "Eng, what sports did you play in high school?"

The Engineer responds proudly, "Baseball, basketball, and football."

"Let me ask you a question. If your football team won the first three games of the season, did your coach say: 'Okay, I think we're good enough. We can stop trying to get better now'?"

The entire group chuckles and then the Eng responds, "Of course not. My football coach was also pushing us hard to improve."

"Exactly right. Even Olympic gold medalists have coaches that push them to get better. In fact, the more potential an athlete has, the harder the coach drives that athlete. That's what you can expect from me. I respect and admire your accomplishments, but we can still improve. Life would be boring if we weren't always striving to improve."

The conversation continues for another hour, and the group enjoys an open and honest discussion about their current strengths and weaknesses. Then they set some short- and long-term goals for each department.

Of course, one round table discussion did not transform this team from mediocre to excellent. There were many others, and the Commanding Officer continued to use the analogy of the coach that pushes those with potential hard so that they can reach that potential. But progress required more than these continued discussions. Incremental improvement and small changes are the items that the Commanding Officer tackled next.

For example, the qualification process on board the ship was lackluster. Training was lackluster and often ineffective. An unacceptable number of people were behind in their qualification progress. These deficiencies weren't lost on the leaders of the ship, but they hadn't been successful in improving them. The Commanding Officer began to implement small tweaks in the training program to breathe some energy into it. For example, he hosted a Jeopardy session of the mechanics against the electricians during one training session—the winner was able to leave the boat one hour early that day. The day before a training session, he spent time with the trainer reviewing the presentation, and he gave him some ideas to spice up the training, such as making the training more scenario-based and more interactive. He focused his energy on the details that he believed if attended to would, in time, snowball into larger and more substantial improvements.

Here are some other examples of small changes that he implemented to propel the crew from mediocre to great.

He made the officer's accountable for the appearance of their staterooms. (On a fast attack submarine there are three staterooms, each housing three officers.) It is not uncommon for the officers' staterooms to degrade into pig sties. Officers are so often inundated with paperwork that their racks (beds) become auxiliary filing cabinets. He directed the Executive Officer to enforce a standard of cleanliness and stowage for sea (meaning no piles of paperwork on officer racks). When this became challenging to enforce he declared two consecutive days as "administration catch up day"—all events were canceled, and the officers were directed to spend these days doing nothing but catching up on their paperwork. This action had a dual positive effect: the staterooms no longer overflowed with stacks of paperwork, and the ship administration was quickly accelerated.

He modernized the wardroom with an upgraded entertainment system, improved artwork on the walls, and demanded that the officers treat the wardroom with great respect in regards to keeping it free of clutter and make it a habit to leave the wardroom slightly better than when they entered. That could be something as simple as brewing a fresh pot of coffee for the oncoming watchsection or tidying up the magazines stowed in the wardroom.

He demanded that all events start exactly on time. The ship had a habit of starting events (training, drill briefs, watchsection turnovers, night work meetings) 5 to 10 minutes late. No more. Great ships start events on time—we are a great ship, and therefore we start events on time. There were indeed growing pains on this as well. A few times the Commanding Officer had the Executive Officer cancel events when they were four minutes into a scheduled event and all the required attendees were still not there. Eventually, events on this Nuclear Submarine started promptly.

As time went on, the Commanding Officer was more confident in implementing change that would lead toward creating a culture of operational excellence. But he displayed patience and recognized the importance of timing—too much at once and he would lose buy-in from the ship's leaders, not enough change and the ship would languish in its mediocrity.

Improving the performance of a Nuclear Submarine is part of the lifecycle of the Nuclear Submarine community. The pendulum is usually swinging towards or away from greatness. In the case where the pendulum is resting at its steady state position, there exists a great opportunity to push the pendulum in the direction toward excellence and, of course, there exists a real risk that unintended consequences could push the pendulum in the wrong direction.

Here are some highlights to consider when inheriting a mediocre organization:

- Mediocre is not a failure. Most teams are average, and many live many decades to tell that tale.
- Observe before you leap into action. It may be tempting to immediately enact a lot of little changes to raise the bar, but doing so may cause resentment, push-back, and unintended negative consequences.
- Openly challenge the leadership team. Use the coach

analogy that even Olympic-grade athletes have coaches who push them harder than they might ask to be.

- Focus on the crew's accomplishments. Nitpicking and criticizing their weaknesses is unlikely to be a successful motivation technique.
- Pick your battles. Choose the items that are most important to you—two or three and go after them relentlessly. The organization expects change when a new leader arrives, but too much change is likely to create resentment and frustration.
- Look for opportunities to create small, incremental, and measurable change. Give others credit when those opportunities snowball into accomplishments.

THE SUPERSTAR TEAM

We saved the most challenging scenario for the last one. You are assigned to an organization, a business unit, or a Nuclear Submarine that is a rock star. Its performance has been stellar for a sustained period; its reputation is impeccable; and your assignment, in theory, is enviable. Unfortunately, organizations rarely sustain superior performance over extended periods of time, unless the organization devotes a tremendous amount of resources to improve its operations. That last sentence may seem illogical. Why would an organization that has achieved operational excellence need to spend resources to enhance its operations? Wouldn't the leaders focus on maintaining their performance? The simple answer is "yes"; that is what most leaders do, and, as a result, their fall from grace is only a matter of time.

If you had an engineering background, you might be inclined, wisely, to reference the Second Law of Thermodynamics that discusses the phenomenon of entropy or disorder, but we won't

former Commanding Officer took command, the ship was one of the worst performing boats in the fleet. The ship's inspection grades were consistently below the fleet average, and the ship's crew struggled to execute planning and maintenance to a satisfactory standard. In those days, the submarine was conducting several critiques a week. In the past 12 months, the leadership convened a total of three critiques.

As we discussed in Chapter 7, there is no one metric to determine what events should be critiqued and which ones should not be. Further, a reasonable person could deduce that a ship that is performing well would have fewer critiques than a ship that is performing poorly. This would be true only if criteria were static. They are not, or at least they shouldn't be. At the heart of the matter, critiques are performed to improve the ship's performance. As the ship's performance increases, it would be reasonable to expect the criteria for conducting a critique to shift. Maybe not to the extent that it continues to perform two critiques a week, but few professional submariners would disagree that if a Nuclear Submarine isn't holding a critique at least once a month, the leadership isn't looking very hard at its performance.

The pendulum is always swinging. Poor performing boats become the great boats and vice versa. There is a natural ebb and flow of talent and performance that occurs aboard Nuclear Submarines. It doesn't take a tremendously great leader to take a poor performing boat and make it a great boat; that happens often. The submarine support structure assures this. Struggling boats are provided the top students from the training pipeline. The same is true with leadership rotations; the submarine support structure will assign the top performers (Department Heads, Chiefs, senior operators and technicians). It's not unlike the poorest performing professional sports teams receiving the highest draft picks. Further, Squadron resources are deployed aggressively to assist the leadership of struggling ships. When you're under the microscope, you will get better . . . and the pendulum swings.

The real measure of a great leader in the Nuclear Submarine community is maintaining operational excellence on a "great" boat. The excellent boats no longer receive first round draft picks. The excellent boats are no longer getting as much support from the Squadron. It takes a great leader to achieve operational excellence, and it takes a tremendously great leader to maintain operational excellence. If the leader of the excellent boat does not take aggressive action, the pendulum will swing in the wrong direction.

There are many different perspectives we could use to evaluate this pendulum swinging phenomenon, but let's stick with the critique process. Every event that is performed by human beings is imperfect. No event occurs on a submarine that is "perfect"; there is always room for improvement. Do we think that Olympic gold medalists train with a standard of attempting to stay as good as they are or with a standard of striving for improvement? Operational excellence requires the same constant and sustained commitment to improvement.

Assume that you could quantify the "imperfection" of an event. Further assume that a leader's critique standard could also be quantified such that she chooses to critique any events above a certain level of imperfection. In that world, the events above the line in Figure 9.1 would be critiqued, and those below the line would not be.

Figure 9.1

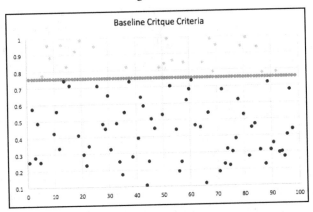

If the critique process is effective and the ship's leadership directs the ship's crew towards operational excellence, we would expect the magnitude of imperfection to reduce over time. Let's assume, the leadership and the crew effort was tremendous such that the boat achieved a level of operational excellence recognized by all. If their critique bar doesn't shift, we might expect that Figure 9.2 represents their magnitude of imperfection.

Figure 9.2

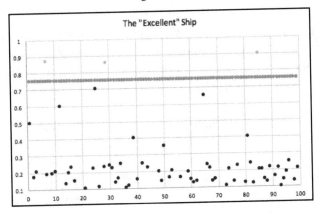

As we see, the ship would have three critiques as only three events exceed the critique criteria bar. However, unless the ship's leadership lowers that critique bar, the number, and magnitude of imperfect events will continue to increase. In fact, even with this "excellent" boat, you can see a few outliers that are approaching the critique bar.

Although we started this train of thought using the term *critique* in its most formal application, as Chapter 7 describes. It is now more appropriate to consider the word *critique* and the associated critique bar on the charts in this section with a less formal connotation, such as any honest analysis about what happened, why it happened, and what can we do to prevent such an event from occurring again. The events that are approaching the critique line in Figure 9.2 represent the "sniff" that the Commanding Officer and the Chief of the Boat have been getting about the ship's operations not being as good as the crew and its leaders believe it to be.

Predictably, if the critique line is not lowered, the imperfect events will continue to increase in magnitude. Perhaps it wouldn't be long before the imperfections in the submarine's operations could be represented by Figure 9.3. The graphic is exaggerated for effect, but we hope you see the point we are driving at. Taking this trend to its logical progression would be equivalent to watching this boat's pendulum swing—hard and fast in the wrong direction.

The leader of an excellent organization, not unlike the coach of a Super Bowl-winning football team, must continue to raise the bar (the equivalent of lowering the bar in these graphics) because there is no standing still in the pursuit of operational excellence. If you are not getting better, then you are getting worse—there are no exceptions to that rule. The second law

Figure 9.3

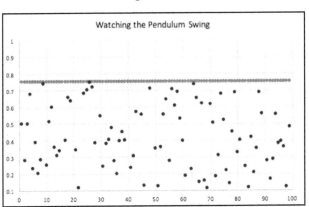

of thermodynamics supports this theory, paraphrased as "the disorder in a system will increase unless energy is spent to prevent this."

Refocusing on our Commanding Officer, his challenge is to take the current command climate (Figure 9.4) and lead the organization represented in Figure 9.5.

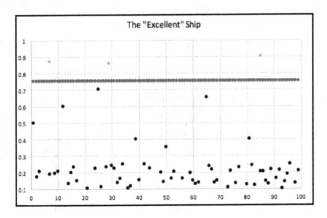

Figure 9.5

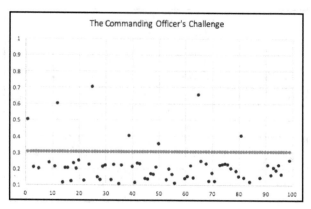

Of course, this is much easier said than done. Almost anyone can increase his level of criticism, but as Theodore Roosevelt reminds us, "It is not the critic who counts." The line in Figures 9.4 and 9.5 does not represent criticism; it represents a standard of brutally honest evaluation and analysis. There is no doubt that to achieve the level of operational excellence that they did, the former Commanding Officer slowly raised the standards (bringing the line of critique criteria down), but then when the leaders start to become the organization's greatest fans, the bar slowly drifts upwards. This is the scenario this Commanding Officer inherited.

Although there were growing pains, obstacles, and bumps in the road, the Commanding Officer in our scenario was able to endure a pendulum swing in the wrong direction but then stopped its momentum and reversed its course back to a place of true operational excellence.

Here are some highlights from the inheriting a superstar team scenario:

- Never believe your own press. As a leader, your job is to continue to strive for improvement. Let others sing your praises, but never sing your own.
- Remember the Rockstar Phenomenon. Sometimes your organization's success is a function of a handful of extremely talented individuals. This is a great situation to find yourself in, but extremely talented individuals tend to move on to bigger and better things more often than not. Your goal in achieving operational excellence should be that the organization and its processes are excellent and do not rely solely on the extraordinary talents of a few folks.
- When you get a sniff that something "isn't right" . . . trust yourself. You did not become a leader accidentally. Your instincts are sharp and should be trusted.
- When you raise the bar in an organization that is performing at an excellent level, you run the risk of becoming the overbearing father that can never be pleased. Respect your team's accomplishments and emphasize that your objective is to coach the team to a level where each individual is operating at his or her full potential.
- Do not be afraid to make personnel changes. Leaders who have contributed to the success of an organization have a tendency to be very defensive and resistant to change. Expect this. However, if after exhausting all of your communication tools, you are unable to achieve a collaborative and compliant relationship, do not live with that dysfunctional relationship.

4. Remind the team how it achieved operational excellence. Discuss the journey and how standards have changed over time. You may be able to reignite the flame of high standards that drove the organization to success.

THE BIG PICTURE

These scenarios were all paper thin. Each deserves a book in its own right because the challenges of entering any leadership role are tremendous. We presented these scenarios to remind you of two things: 1) Leadership has no one-size-fits-all solution. 2) Always be aware of the pendulum—many leaders focus too much attention on where an organization "is" instead of where the organization is "trending." Assessing the status of an organization is challenging, but many people can accomplish this. The information that very few people are adept at ascertaining is where an organization is trending. For organizations that have bottom line financial standards, recognize that financial performance is a lagging indicator—it is giving you information about the past. The best leaders identify metrics in their organizations that are leading indicators so they can best adjust as necessary to ensure future performance is maximized.

CHAPTER 10

HOT WASH

When an evolution is completed, often the leader will call for a "hot wash"—a quick review of the performance of the event. The process is typically a rapid, direct, and high-level discussion to ensure that everyone has the opportunity to provide input or ask questions about the evolution. Additionally, it ensures that lessons learned or action items are disseminated rapidly to the group.

Let's hot wash "Extreme Operational Excellence."

Often the information that we obtain after reading a book fades from memory. This chapter is provided as a book summary should you ever need to review the highlights of the topics discussed. For ease of future reference, this chapter is bare-bones and succinct. If you are a "read the first and last chapter" type of reader, you may be left with less than you expect, but this was a deliberate trade-off to keep the chapter short and sweet.

OPERATIONAL EXCELLENCE

We define Operational Excellence as a culture of knowledge, brutally honest self-assessment, continuous improvement, and intellectual integrity.

The United States Nuclear Navy is highly regarded as an organization that is fully dedicated to the pursuit of Operational Excellence. Further, the United States Nuclear Navy has succeeded in avoiding catastrophes in an environment where normal accidents can be expected due to risk factors and complexity. Given this definition and this statement about the Nuclear Navy, the obvious question is: Why? Why does this take on Operational Excellence make the Nuclear Submarine community a model worthy of emulation by other organizations? Why does this take on Operational Excellence make the performance of the Nuclear Submarine community a model worthy of emulation by other organizations?

The Nuclear Navy's journey to Operational Excellence is rooted in its culture. That culture is characterized by five organizational values:

1. Knowledge and Learning

2. Procedural Compliance

3. Questioning Attitude

4. Watchteam Backup

5. Integrity

These five organizational values do not exist independently of one another. They interact and rely on each other. We will look at each individually but recognize that a degradation in one area will manifest in a breakdown across the others.

KNOWLEDGE AND LEARNING

An organization that does not invest in providing and improving the knowledge of its operators regarding complex systems is asking for trouble. In the U.S. Nuclear Navy, the focus on training warrants a huge investment in time and resources before the operator is even sent to the operational fleet. Then the operator on a nuclear warship goes through more extensive training and a robust qualification process. Civilian organizations may take a different strategy by hiring qualified operators rather than training their own in-house, but, either way, operators must have a deep knowledge of the equipment and system interoperability.

Equally important, is the organization's commitment to learning. Leaders of an organization must develop an environment that encourages its members to continuously seek out new knowledge and resist the "we have always done it that way" mentality. Much of this learning will come from taking a hard look at mistakes and not shrinking from the difficult work of identifying root causes and taking corrective action.

PROCEDURAL COMPLIANCE

The Nuclear Submarine community's commitment to procedural compliance occasionally draws criticism from members of the civilian power generation industry, as well as other military communities. The criticism often infers that rigid allegiance to procedures creates thoughtless and near-robotic operators. We steadfastly defend the Nuclear Submarine community on this front. Remember that all the five cultural values must work in concert to create an environment of Operational Excellence. Without a questioning attitude and the desire to continuously learn more, the benefits of procedural compliance degrade quickly. Thoughtful operators will consider their commitment to procedural compliance as a reflection of their respect for the organization's processes and ability to feed lessons learned into those processes.

Additionally, thoughtful and knowledgeable operators recognize when systems are operating outside of the parameters that the procedures assume. No procedure can be written to cover all initial conditions and system lineups. In those situations, the knowledgeable and experienced operator is expected to deviate from the procedure after deliberate and thoughtful discussion with supervisors. In the event of casualties, the casualty procedures allow the operator to quickly react with "muscle memory" rather than have to deliberate over the right course of action to stabilize the plant.

QUESTIONING ATTITUDE

To some, creating an environment that encourages a Questioning Attitude throughout an organization may sound like a recipe for chaos. There is no doubt that on its own and improperly led, this environment can be annoying at best and completely disruptive at worst. The Questioning Attitude environment that serves as a pillar of the Nuclear Submarine community's pursuit of Operational Excellence is more about mindset and less about actually actually asking questions. When combined with a high level of knowledge, integrity, and a respect for procedural compliance, engaged minds become an organization's most coveted asset. An engaged mind is precisely the objective of building an Questing Attitude environment. An engaged mind is in a perennial state mindfulness, always asking: What am I doing now? Is there a more efficient way of doing this? Am I prepared to deal with the consequences of what I am doing? Is there an unintended consequence of what I am doing that I haven't identified? Who else will be impacted by what I am doing? What about that other person—why is he doing what he is doing? Do I understand what he is doing? Should I? These simple questions, sometimes from the least-experienced personnel, can save the day or spark the organization's next great idea .

It takes constant leadership feedback at all levels to establish the optimal balance between questioning everything and questioning nothing. Establishing this balance is extraordinarily challenging, but the results can be immeasurably positive. A leader has to be intentional in his or her actions and communications each and every moment as the organization will take its cue from the leader. This is not a requirement of just the top leader but of leaders at all levels. Front-line supervisors, especially in their first assignment, need guidance and mentoring in this. We have observed too often that new leaders and especially new front-line managers struggle with balancing the questioning attitude from operators who were just previously their peers. Senior leaders should not neglect this important cultural value.

WATCHTEAM BACKUP

Watchteam Backup is the action side of an engaged employee. While Questioning Attitude is the thinking part, people have to be willing to take action to step in and do something when they believe operations are going awry. How many incident investigations discover testimony stating: "I didn't think things were going right, but I didn't say anything"? Sometimes, Watchteam Backup requires people to step across their normal "lanes of operation" and into "other people's business." The stovepipes present in many organizations create barriers to these Watchteam Backup actions. As a senior leader, you must look for opportunities to encourage cross-functional interaction so that your teams are comfortable backing up each other. You also must be willing to listen patiently to a question from a junior employee. Of course, there are times when you can't explain why you need something done due to time constraints. Trust us, we understand. That's what makes this even more challenging. And please don't think that we are advocating for prolonged decision-making-by-committee that has paralyzed so many organizations whose leaders seek to get input from everyone on everything. That type of culture is not what we seek. The journey towards Operational Excellence requires finding the right balance throughout the organization to allow for both efficiency of action and patience without becoming overly bound to either extreme.

INTEGRITY

"Doing the right thing when no one is watching" is our working definition of Integrity. It's a simple phrase, but we all know it's hard to do. This is not a book on philosophy, ethics, or human psychology. But, mature adults understand that when no one is watching, humans have a tendency to take short-cuts, cover up mistakes, take the easy path, or otherwise do the wrong thing. We realize that this is a pretty negative outlook on the human condition. But, over two decades of operating complex machines with young sailors and officers leads us to default to this outlook so that we intentionally put effort into instilling integrity throughout the organization.

If you study major catastrophes across many industries, we believe you will come to the same conclusion that a high level of organizational integrity is not a natural outcome without a strong, sustained, and persistent leadership effort. Leaders at all levels must be aligned in their pursuit of maintaining the highest standards of organization integrity. If mid-level leaders "look the other way" to get results that the senior leadership is expecting, instead of communicating and addressing their concerns and observations, major incidents can occur. This requires a long view of success and a willingness to support all levels of the chain of command when things go wrong.

THE FEEDBACK LOOP

These five cultural values are hard to argue with. Who wouldn't want these attributes in their organization? But, having them is not as simple as talking about these issues occasionally and putting up a few posters on the break room bulletin board. So here is the hard part—critiquing the events when things go badly. The Nuclear Navy critique process, which includes the gathering of evidence, identification of root causes, and assignment and execution of corrective actions, is the feedback loop to the culture that leaders must be involved in. As a former Commanding Officer and an Engineer Officer of Nuclear Submarines, we have many, many critiques under our belts. Neither of us look back on those critiques as good times. Done right, they will be painful to a degree. Done wrong, they will be even more painful, as the organization will not learn the lessons it should and the next incident will be worse.

One of our observations since we left the Navy is that operational leaders in many organizations turn these critique activities over to an administrative support team sometimes called "Corrective Action Team" or "Continuous Improvement Team." While we both fully understand that the Chief Executive Officer of a large organization cannot attend every incident critique, we assure you that the Nuclear Navy's expectation is that the Commanding Officer of a billion-dollar warship is expected to personally be involved in the critique process. You will have to consider your company's organizational structure and size, but we encourage you to develop an expectation that a senior operational leader is personally involved.

Admiral Rickover explained in the speech excerpt, that we provided in the introduction, that he expected a high level of personal involvement by leaders in the Nuclear Navy. He set the example with his own involvement and attention to the details. Commanding Officers of Nuclear Submarines and Aircraft Carriers have to write personal letters to the Director of Naval Reactors when an incident occurs. Others are copied in on the letters, but they go directly from the Commanding Officer to the Four Star Admiral. How many organizations have a policy that requires the Plant Manager to send a personal letter explaining what happened in plain language when an incident occurs even if the incident does not include loss of life, injury, or a high dollar cost? Yes, you have to establish a reasonable policy of when these incident reports are sent, or the senior leadership will be inundated with low-level incident reports. But the bar is set purposely low in the Nuclear Navy so that small incidents are brought to the Commanding Officer's and the Four Star Admiral's attention. This alone is a critical factor in the Nuclear Navy's culture of Operational Excellence. Ignore this at your organization's peril.

LEADERSHIP

We can't overemphasize the high level of quality leadership that is needed to develop and maintain a culture of Operational Excellence. Strong leadership is not a program. It's a cultural way of life that takes time and energy to develop. In Chapter 9 we described the concept of Pendulum Leadership. We have observed over many years on many Nuclear Submarines that when leaders blindly apply these principles, they sometimes do more damage than good. If the operators have poor knowledge of the systems but leaders try to emphasize Questioning Attitude, chaos may result and operations will degrade. Likewise, if Integrity is weak and Procedural Compliance is not a priority, focusing on training will not be as useful, and a major incident is likely just around the corner. Leaders should first evaluate where their organization stands against the cultural principles desired. Don't use results as the sole indicator of where you sit on the Operational Excellence spectrum. As discussed in Chapter 1, results alone may be a result of market situations that have changed or team members and leaders who have moved on. Results are a lagging indicator.

Measuring the culture is not an easy thing. The Nuclear Navy dictates that a new Commanding Officer and Engineer Officer are required to have a 30-day observation period to evaluate the culture of their new organization. There is also a 90-day overlap required between the Commanding Officer, Executive Officer, and Engineer Officer transitions (except in cases of emergency). These overlaps allow for continuity among the top three leaders on a submarine. These time restraints are designed to reduce the risks that rapid change in a highly complex environment can invite.

You should consider the wisdom of this organizational constraint. If you are a new CEO or General Manager and you want to implement some of the ideas learned in this book, we urge you to be deliberate about establishing where you are today and where you want to be, and then map out a plan to get there. If you are new, you may be able to make these observations yourself. If you have been in the job for a while and these concepts are new to you, you may want to bring in someone from outside your specific business unit to help you identify issues. A common piece of wisdom for new Commanding Officers is that new leaders have about six months to fix problems before they become part of the problem. Indeed, once you are part of the organization for a period of time, it is very hard to see the problems as clearly.

Leading towards Operational Excellence is not easy. It will take an intentional and sustained effort through communication, trust, personal observation, hand wringing incident evaluations, direct feedback to operators, personal integrity, and energy. This book barely touches on the leadership skills required at all levels to achieve the sustained journey towards Operational Excellence in a highly complex and dangerous environment such as a Nuclear Submarine. As we explained in the introduction, we are merely the benefactors of "growing up" in this culture of excellence established by Admiral Rickover and his team over six decades ago. These are not our new ideas. We don't claim to be perfect leaders in these areas. We merely want to share with you the lessons we've learned about how the U.S. Nuclear Navy works and performs so well.

We sincerely hope that this book has helped you in your journey towards Operational Excellence. If our efforts have averted one safety mishap or helped an organization reach its goals, our work has been worthwhile. We look forward to hearing your "sea stories."